Raphael Meldola

Coal And What We Get From It

A Romance of Applied Science

Raphael Meldola

Coal And What We Get From It
A Romance of Applied Science

ISBN/EAN: 9783744674881

Printed in Europe, USA, Canada, Australia, Japan

Cover: Foto ©Thomas Meinert / pixelio.de

More available books at **www.hansebooks.com**

COAL;

AND WHAT WE GET FROM IT.

THE ROMANCE OF SCIENCE.

COAL
AND WHAT WE GET FROM IT.

A Romance of Applied Science.

EXPANDED FROM THE NOTES OF A LECTURE
DELIVERED IN THE THEATRE OF THE
LONDON INSTITUTION, JAN. 20TH, 1890.

BY

RAPHAEL MELDOLA, F.R.S., F.I.C., &c.,
PROFESSOR OF CHEMISTRY IN THE FINSBURY TECHNICAL COLLEGE,
CITY AND GUILDS OF LONDON INSTITUTE.

PUBLISHED UNDER THE DIRECTION OF THE COMMITTEE
OF GENERAL LITERATURE AND EDUCATION APPOINTED BY THE
SOCIETY FOR PROMOTING CHRISTIAN KNOWLEDGE.

LONDON:
SOCIETY FOR PROMOTING CHRISTIAN KNOWLEDGE,
NORTHUMBERLAND AVENUE, CHARING CROSS, W.C.;
43, QUEEN VICTORIA STREET, E.C.
BRIGHTON: 135, NORTH STREET. NEW YORK: E. & J. B. YOUNG & CO.

1891.

TO
WILLIAM HENRY PERKIN,
Ph.D., F.R.S.,

THE FOUNDER OF THE COAL-TAR COLOUR INDUSTRY,
THIS BOOK IS DEDICATED BY THE AUTHOR.

PREFACE.

THIS is neither a technical manual, nor a treatise dealing with the history of a particular branch of applied science, but it partakes somewhat of the character of both. It is an attempt—perhaps somewhat bold—to present in a popular form an account of the great industry which has arisen out of the waste from the gas-works. In the strictest sense it is a romance of dirt. To render intelligible the various stages in the evolution of the industry, without assuming any knowledge of chemical science on the part of the general reader, has by no means been an easy task, and I have great misgivings as to the success of my effort. But there is so much misapprehension concerning the history and the mode of production of colouring-matters from coal-tar, that any attempt to strip the industry of its mystery in this, the land of its birth, cannot but find justification. Although the theme is a favourite one with popular lecturers, it is generally treated in a superficial way, leaving the audience only in possession of the bare fact that dyestuffs, &c., have by some means or other been obtained from coal-tar. I have endeavoured to go somewhat beyond this, and to give some notion of the scientific principles underlying the subject. If the reader can follow these pages, in which not a chemical formula appears, with the same interest and with the same desire to know more about the subject that was manifested by the audience at the London Institution, before whom the lecture was delivered, my object will have been accomplished. To the Board of Managers of that Institution my thanks are due for the opportunity which they have afforded me of attempting to extend that popular knowledge of applied science for which there is such a healthy craving in the public mind at the present time.

R. M.

6 *Brunswick Square, W.C.*

CONTENTS.

CHAPTER I.

Origin of Coal, 9. Coal of various ages, 11. Graphite, 12. Recent Vegetable Deposits, 13. Mode of occurrence of Coal, 13. Structure of Coal, 15. Uses of Coal, 16. Coal a source of Energy, 17. Mechanical Equivalent of Heat, 19. Value of Coal as a Fuel, 20. Small efficiency of Steam-engines, 21. Mechanical value of Coal, 22. Whence Coal derives its Energy, 22. Chemical Composition of Coal, 23. Growth of Plants, 26. Solar Energy, 28. Transformation of Wood into Coal, 30. Destructive Distillation of Coal, 33. Experiments of Becher, 34; of Dean Clayton, 35; of Stephen Hales, 37; of Bishop Watson, 37; of the Earl of Dundonald, 39. Coal-gas introduced by Murdoch, 40. Spread of the new Illuminant, 41. Manufacture of Coal-gas, 42. Quantitative results, 45. Uses of Coke, 47. Goethe's visit to Stauf, 48. Bishop Watson on waste from Coke-ovens, 50. Shale-oil Industry, 50. History of Coal-mining, 57. Introduction of Coal into London, 58. The Coal resources of the United Kingdom, 60. Competition between Electricity and Coal-gas, 62.

CHAPTER II.

Ammoniacal Liquor of Gas-works, 64. Origin of the Ammonia, 65. Ammonia as a Fertilizer, 65. Other uses of Ammonia, 67. Annual production of Ammonia, 68. Utilization of Coal-tar,

69. The Creosoting of Timber, 70. Early uses of the Light Tar Oils, 71. Discovery of Benzene by Faraday; isolation from Tar Oil by Hofmann and Mansfield, 73. Discovery of Mauve by Perkin, 74. History of Aniline, 75. The Distillation of Coal-tar, 77. Separation of the Hydrocarbons of the Benzene Series, 82. Manufacture of Aniline and Toluidine, 87. History and Manufacture of Magenta, 89. Blue, Violet, and Green Dyes from Magenta, 92. The Triphenylmethane Group, 97. The Azines, 108. Lauth's Violet and Methylene Blue, 111. Aniline Black, 114. Introduction of Azo-dyes, 115. Aniline Yellow, Manchester Brown, and Chrysoïdine, 118. The Indulines, 121. Chronological Summary, 122.

CHAPTER III.

Natural Sources of Indigo, 124. Syntheses of the Colouring-matter, 126. Carbolic Oil, its treatment and its constituents, 129. Phenol Dyes, 132. Salicylic Acid and its uses, 134. Picric Acid, 136. Naphthalene and its applications, 139. The Albo-carbon Light, 140. Phthalic Acid and the Phthaleïns, 145. Magdala Red, 149. Azo-dyes from the Naphthols, Naphthylamines, and their Sulpho-acids, 150. Naphthol Green, the Oxazines, and the Indophenols, 161. Creosote Oil, 163. The Lucigen Burner, 163. Anthracene Oil, 167. The Discovery of Artificial Alizarin, and its effects on Madder growing, 167. The industrial isolation of Anthracene and its conversion into Colouring-matters, 171. Pitch, and its uses, 176. Patent Fuel, or Briquettes, 178. Coal-tar products in Pharmacy, 178. Aromatic Perfumes, 185. Coal-tar Saccharin, 186. Coal-tar Products in Photography, 188. Coal-tar Products in Biology, 192. Value of the Coal-tar Industry, 194. The Coal-tar Industry in relation to pure Science, 196. Permanence of the Artificial Colouring-matters, 198. Chronological Summary, 200. Addendum, 202.

COAL;
AND WHAT WE GET FROM IT.

CHAPTER I.

"Hier [1771] fand sich eine zusammenhängende Ofenreihe, wo Steinkohlen abgeschwefelt und zum Gebrauch bei Eisenwerken tauglich gemacht werden sollten; allein zu gleicher Zeit wollte man Oel und Harz auch zu Gute machen, ja sogar den Russ nicht missen, und so unterlag den vielfachen Absichten alles zusammen."— Goethe, *Wahrheit und Dichtung*, Book X.

To get at the origin of the familiar fuel which blazes in our grates with such lavish waste of heat, and pollutes the atmosphere of our towns with its unconsumed particles, we must in imagination travel backwards through the course of time to a very remote period of the world's history. Ages before man, or the species of animals and plants which are contemporaneous with him, had appeared upon the globe, there flourished a vegetation not only remarkable for its luxuriance, but also for the circumstance that it consisted to a preponderating extent of non-flowering or cryptogamic plants. In swampy areas, such as the deltas at the mouths of great rivers, or in shallow lagoons bordering a coast margin, the jungles of ferns and tree-ferns, club-mosses and horse-tails, sedges, grasses, &c., grew and died down year by year, forming a consolidated mass of vegetable matter much in the same way that a peat bed or a mangrove swamp is accumulating organic deposits at the present time. In the course of geological change these beds of compressed vegetation became gradually depressed, so that marine or fresh-water sediment was deposited over them, and then once more the vegetation spread and flourished to furnish another accumulation of vegetable matter, which in its turn became submerged and buried under sediment, and so on in successive alternations of organic and sedimentary deposits.

But these conditions of climate, and the distribution of land and water favourable to the accumulation of large deposits of vegetable matter, gradually gave way to a new order of things. The animals and plants adapted to the particular conditions of existence described above gave rise to descendants modified to meet the new conditions of life. Enormous thicknesses of other deposits were laid down over the beds of vegetable remains and their intercalated strata of clay, shale, sandstone, and limestone. The chapter of the earth's history thus sealed up and stowed away among her geological records relates to a period now known as the Carboniferous, because of the prevalence of seams or beds of coal throughout the formation at certain levels. By the slow process of chemical decomposition without access of air, modified also by the mechanical pressure of superincumbent formations, the vegetable deposits accumulated in the manner described have, in the lapse of ages, become transformed into the substance now familiar to us as coal.

Although coal is thus essentially a product of Carboniferous age, it must not be concluded that this mineral is found in no other geological formation. The conditions favourable for the deposition of beds of vegetable matter have prevailed again and again, at various periods of geological time and on different parts of the earth, although there is at present no distinct evidence that such a luxuriant growth of vegetation, combined with the other necessary conditions, has ever existed at any other period in the history of the globe. Thus in the very oldest rocks of Canada and the northern States of America, in strata which take us back to the dawn of geological history, there is found abundance of the mineral graphite, the substance from which black-lead pencils are made, which is almost pure carbon. Now most geologists admit that graphite represents the carbon which formed part of the woody tissue of plants that lived during those remote times, so that this mineral represents coal in the ultimate stage of carbonization. In some few instances true coal has been found converted into graphite *in situ* by the intrusion of veins of volcanic rock (basalt), so that the connection between the two minerals is more than a mere matter of surmise.

Then again we have coal of pre-Carboniferous age in the Old Red Sandstone of Scotland, this being of course younger in point of time than the graphite of the Archæan rocks. Coal of post-Carboniferous date is found in beds of Permian age in Bavaria, of Triassic age in Germany, in the Inferior Oolite of Yorkshire belonging to the Jurassic period, and in the Lower Cretaceous deposits of north-western Germany. Coming down to more recent geological periods, we have a coal seam of over thirty feet in thickness in the northern Tyrol of Eocene age; we have brown coal deposits of Oligocene age in Belgium and Austria, and, most remarkable of all, coal has been found of Miocene, that is, mid-Tertiary age, in the Arctic regions of Greenland within a few degrees of the North Pole. Thus the formation of coal appears to have been going on in one area or another ever since vegetable life appeared on the globe, and in the peat bogs, delta jungles, and mangrove swamps of the present time we may be said to have the deposition of potential coal deposits for future ages now going on.

Although in some parts of the world coal seams of pre-Carboniferous age often reach the dignity of workable thickness, the coal worked in this country is entirely of Carboniferous date. After the explanation of the mode of formation of coal which has been given, the phenomena presented by a section through any of our coal measures will be readily intelligible (see Fig. 1). We find seams of coal separated by beds of sandstone, limestone, or shale representing the encroachment of the sea and the deposition of marine or estuarine sediment over the beds of vegetable remains. The seams of coal, varying in thickness from a few inches to three or four feet, always rest on a bed of clay, known technically as the "underclay," which represents the soil on which the plants originally grew. In some instances the seams of coal with their thin "partings" of clay reach an aggregate thickness of twenty to thirty feet. In many cases the very roots of the trees are found upright in a fossilized condition in the underclay, and can be traced upwards into the overlying coal beds; or the completely

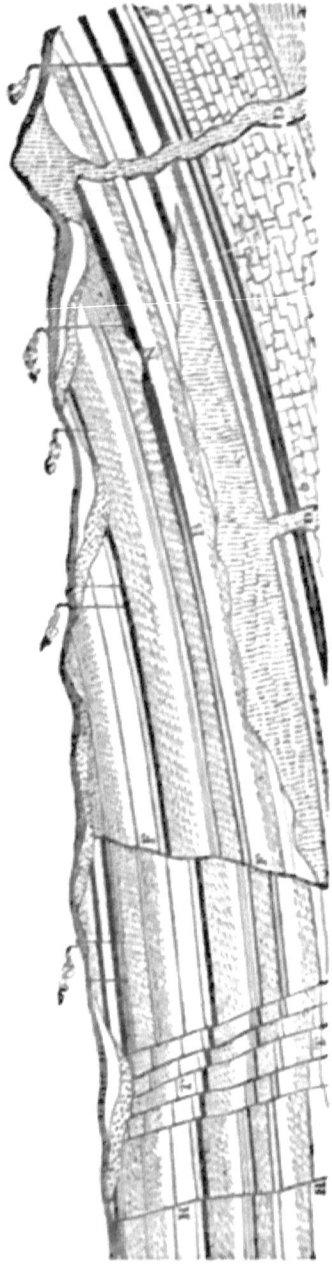

FIG. 1.—Section through Carboniferous strata showing seams of coal. Dislocations, or "faults," so common in the Coal Measures, are shown at H, T, and F. Intrusions of igneous rock are shown at D. At B is shown the coalescence of two seams, and at N the local thinning of the seam. The vertical lines indicate the shafts of coal mines.

carbonized trunk is found erect in the position in which the tree lived and died (see Fig. 2).

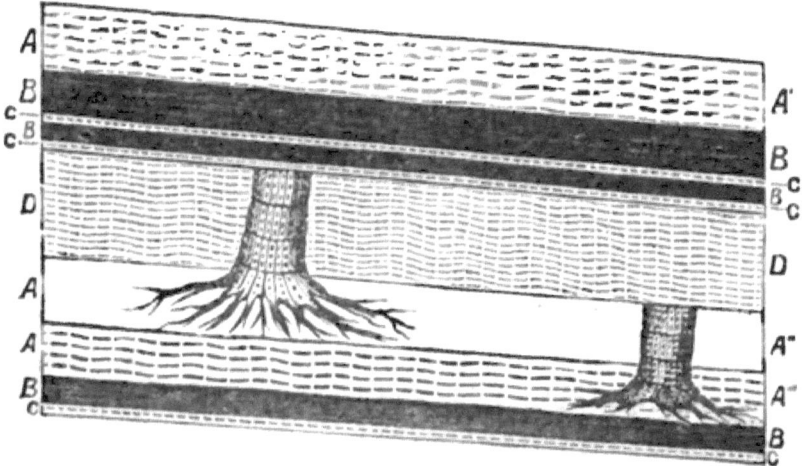

FIG. 2.—Section showing coal seams and upright trunks attached to roots *in situ*. A', A", A''', beds of shale. B, coal seams. C, underclay. D, sandstone.

Owing to the chemical and mechanical forces to which the original vegetable deposit has been subjected, the organic structure of coal has for the most part been lost. Occasionally, however, portions of leaves, stems, and the structure of woody fibre can be detected, and thin sections often show the presence of spore-cases of club-mosses in such numbers that certain kinds of coal appear to be entirely composed of such remains. But although coal itself now furnishes but little direct evidence of its vegetable origin, the interstratified clays, shales, and other deposits often abound with fossilized plant remains in every state of preservation, from the most delicate fern frond to the prostrate tree trunk many yards in length. It is from such evidence that our knowledge of the Carboniferous flora has been chiefly derived.

Now this carbonized vegetation of a past age, the history of which has been briefly sketched in the foregoing pages, is one of the chief sources of our industrial supremacy as a nation. We use it as fuel for generating the steam which drives our engines, or for the production of heat wherever heat is wanted. In metallurgical operations we consume enormous quantities of coal for extracting metals from their ores, this consumption being especially great in the case of iron smelting. For this last operation some kinds of raw coal are unsuitable, and such coal is converted into coke before being used in the blast furnace. The fact that the iron ore and the coal occur in the same district is another cause of our high rank as a manufacturing nation.

It has often been a matter of wonder that iron ore and the material essential for extracting the metal from it should be found associated together, but it is most likely that this combination of circumstances, which has been so fortunate for our industrial prosperity, is not a mere matter of accident, but the result of cause and effect. It is, in fact, probable that the iron ore owes its origin to the reduction and precipitation of iron compounds by the decomposing vegetation of the Carboniferous period, and this would account for the occurrence of the bands of ironstone in the same deposits with the coal. In former times, when the area in the south-east of England known as the Weald was thickly wooded, the towns and villages of this district were the chief centres of the iron manufacture. The ore, which was of a different kind to that found in the coal-fields, was smelted by means of the charcoal obtained from the wood of the Wealden forests, and the manufacture lingered on in Kent, Sussex, and Surrey till late in the last century, the railings round St. Paul's, London, being made from the last of the Sussex iron. When the northern coal-fields came to be extensively worked, and ironstone was found so conveniently at hand, the Wealden iron manufacture declined, and in many places in the district we now find disused furnaces and heaps of buried slag as the last witnesses of an extinct industry.

From coal we not only get mechanical work when we burn it to generate heat under a steam boiler, but we also get chemical work out of it when we employ it to reduce a metallic ore, or when we make use of it as a source of carbon in the manufacture of certain chemical products, such as the alkalies. We have therefore in coal a substance which supplies us with the power of doing work, either mechanical, chemical, or some other form, and anything which does this is said to be a source of energy. It is a familiar doctrine of modern science that energy, like matter, is indestructible. The different forms of energy can be converted into one another, such, for example, as chemical energy into heat or electricity, heat into mechanical work or electricity, electricity into heat, and so forth, but the relationship between these convertible forms is fixed and invariable. From a given quantity of chemical energy represented, let us say, by a certain weight of coal, we can get a certain fixed amount of heat and no more. We can employ that heat to work a steam-engine, which we can in turn use as a source of electricity by causing it to drive a dynamo-machine. Then this doctrine of science teaches us that our given weight of coal in burning evolves a quantity of heat which is the equivalent of the chemical energy which it contains, and that this quantity of heat has also its equivalent in mechanical work or in electricity. This great principle—known as the Conservation of Energy—has been gradually established by the joint labours of many philosophers from the time of Newton downwards, and foremost among these must be ranked the late James Prescott Joule, who was the first to measure accurately the exact amount of work corresponding to a given quantity of heat.

In measuring heat (as distinguished from temperature) it is customary to take as a unit the quantity necessary to raise a given weight of water from one specified temperature to another. In measuring work, it is customary to take as a unit the amount necessary to raise a certain weight at a specified place to a certain height against the force of gravity at that place. Joule's unit of heat is the quantity necessary to raise one pound of water from 60° to 61° F., and his unit of work is the foot-pound, *i.e.* the quantity necessary to raise a weight of one pound to a height of one foot. Now the quantitative relationship between heat and work measured by Joule is expressed by saying that the mechanical equivalent of heat is about 772 foot-pounds, which means that the quantity of heat that would raise one pound of water 1° F. would, if converted into work, be capable of raising a one-pound weight to a height of 772 feet, or a weight of 772 lbs. to a height of one foot.

This mechanical equivalent ought to tell us exactly how much power is obtainable from a certain weight of coal if we measure the quantity of heat given out when it is completely burnt.

Thus an average Lancashire coal is said to have a calorific power of 13,890, which means that 1 lb. of such coal on complete combustion would raise 13,890 lbs. of water through a temperature of 1° F., if we could collect all the heat generated and apply it to this purpose. But if we express this quantity of heat in its mechanical equivalent, and suppose that we could get the corresponding quantity of work out of our pound of coal, we should be grievously mistaken. For in the first place, we could not collect all the heat given out, because a great deal is communicated to the products of combustion by which it is absorbed, and locked up in a form that renders it incapable of measurement by our thermometers. In the next place, if we make an allowance for the quantity of heat which thus disappears, even then the corrected calorific power converted into its mechanical equivalent would not express the quantity of work practically obtainable from the coal.

In the most perfectly constructed engine the whole amount of heat generated by the combustion of the coal is not available for heating the boiler—a certain quantity is lost by radiation, by heating the material of the furnace, &c., by being carried away by the products of combustion and in other ways. Moreover, some of the coal escapes combustion by being allowed to go away as smoke, or by remaining as cinders. Then again, in the engine itself a good deal of heat is lost through various channels, and much of the working power is frittered away through friction, which reconverts the mechanical power into its equivalent in heat, only this heat is not available for further work, and is thus lost so far as the efficiency of the engine is concerned. These sources of loss are for the most part unavoidable, and are incidental to the necessary imperfections of our mechanism. But even with the most perfectly conceivable constructed engine it has been proved that we can only expect one-sixth of the total energy of the fuel to appear in the form of work, and in a very good steam-engine of the present time we only realize in the form of useful work about one-tenth of the whole quantity of energy contained in the coal. Although steam power is one of the most useful agencies that science has placed at the disposal of man, it is not generally recognized by the uninitiated how wasteful we are of Nature's resources. One of the greatest problems of applied science yet to be solved is the conversion of the energy latent in coal or other fuel into a quantity of useful work approximating to the mechanical equivalent much more closely than has hitherto been accomplished.

But although we only get this small fraction of

the whole working capability out of coal, the actual amount of energy dormant in this substance cannot but strike us as being prodigious. It has already been said that a pound of coal on complete combustion gives out 13,890 heat units. This quantity of heat corresponds to over 10,000,000 foot-pounds of work. A horse-power may be considered as corresponding to 550 foot-pounds of work per second, or 1,980,000 foot-pounds per hour. Thus our pound of coal contains a store of energy which, if capable of being completely converted into work without loss, would in one hour do the work of about five and a half horses. The strangest tales of necromancy can hardly be so startling as these sober figures when introduced for the first time to those unaccustomed to consider the stupendous powers of Nature.

If energy is indestructible, we have a right to inquire in the next place from whence the coal has derived this enormous store. A consideration of the origin of coal, and of its chemical composition, will enable this question to be answered. The origin of coal has already been discussed. Chemically considered, it consists chiefly of carbon together with smaller quantities of hydrogen, oxygen, and nitrogen, and a certain amount of mineral matter which is left as ash when the coal is burnt. The following average analyses of different varieties will give an idea of its chemical composition :—

Variety of Coal.	Carbon.	Hydrogen.	Oxygen.	Nitrogen.	Ash.
S. Staffordshire	73·4	5·0	11·7	1·7	2·3
Newcastle (Caking)	80·0	5·3	10·7	2·2	1·7
Cannel (Wigan)	81·2	5·6	7·9	2·1	2·5
Anthracite (Welsh)	90·1	3·2	2·5	0·8	1·6

There are in addition to these constituents small quantities of sulphur and a certain variable amount of water (5 to 10 per cent.) in all coals, but the elements which most concern us are those heading the respective columns.

From the foregoing analyses, which express the percentage composition, it will be seen that carbon is by far the most important constituent of coal. Carbon is a chemical element which is found in a crystalline form in nature as the diamond, and which forms a most important constituent of all living matter, whether animal or vegetable. Woody fibre contains a large quantity of this element, and the carbon of coal is thus accounted for; it was accumulated during the growth of the plants of the Carboniferous period.

Now carbon is one of those elementary substances which are said to be *combustible*, which means that if we heat it in atmospheric air it gives out heat and light, and gradually disappears, or, as we say, burns away. The heat which is given out during combustion represents the chemical energy stored up in the combustible, for combustion is in fact the chemical union of one substance with another with the development of heat and light. When carbon burns in air, therefore, a chemical combination takes place, the air supplying the other substance with which the carbon combines. That other substance is also an element—it is the invisible gas which chemists call oxygen, and which forms one-fifth of the bulk of atmospheric air, the remainder consisting of the gas nitrogen and small quantities of other gases with which we shall have more to do subsequently. When oxygen and carbon unite under the conditions described, the product is an invisible gas known as carbon dioxide, and it is because this gas is invisible that the carbon seems to disappear altogether on combustion. In reality, however, the carbon is not lost, for matter is as indestructible as energy, but it is converted into the dioxide which escapes as gas under ordinary circumstances. If, however, we burn a given weight of carbon with free access of air, and collect the product of combustion and weigh it, we shall find that the product weighs more than the carbon, by an amount which represents the weight of oxygen with which the element has combined. By careful experiment it would be found that one part by weight of carbon would give three and two-third parts by weight of carbon dioxide. . If, moreover, we could measure the quantity of heat given out by the complete combustion of one pound of carbon, it would be found that this quantity would raise 14,544 lbs. of water through 1° F., a quantity of heat corresponding to over eleven million foot-pounds of work, or about seven and three-quarters horse-power per hour.

Here then is the chief source of the energy of coal—the carbon of the plants which lived on this earth long ages ago has lain buried in the earth, and when we ignite a coal fire this carbon combines with atmospheric oxygen, and restores some of the energy that was stored up at that remote period. But the whole of the energy dormant in coal is not due to the carbon, for this fuel contains another combustible element, hydrogen, which is also a gas when in the free state, and which is one of the constituents of water, the other constituent being oxygen. In fact, there is more latent energy in hydrogen, weight for weight, than there is in carbon, for one pound of hydrogen on complete combustion would give enough heat to raise 62,032 lbs. of water through 1° F. Hydrogen in burning combines with oxygen to form water, so that the products of the complete combustion of coal are carbon dioxide and water. The amount of heat contributed by the hydrogen of coal is, however, comparatively insignificant, because there is only a small percentage of this element present, and we thus come to the conclusion that nearly all the work that is done by our steam-engines of the present time is drawn from the latent energy of the carbon of the fossilized vegetation of the Carboniferous period.

The conclusion to which we have now been led leaves us with the question as to the *origin* of the energy of coal still unanswered. We shall have to go a step further before this part of our story is complete, and we must form some kind of idea of the way in which a plant grows. Carbon being the chief source of energy in coal, we may for the present confine ourselves to this element, of which woody fibre contains about 50 per cent. Consider the enormous gain in weight during the growth of a plant; compare the acorn, weighing a few grains, with the oak, weighing many tons, which arises from it after centuries of growth. If matter is indestructible, and never comes into existence spontaneously, where does all this carbon come from? It is a matter of common knowledge that the carbon of plants is supplied by the atmosphere in the form of carbon dioxide—the gas which has already been referred to as resulting from the combustion of carbon. This gas exists in the atmosphere in small quantity—about four volumes in 10,000 volumes of air; but insignificant as this may appear, it is all important for the life of plants, since it is from this source that they derive their carbon. The origin of the carbon dioxide, which is present as a normal constituent of the atmosphere, does not directly concern us at present, but it is important to bear in mind that this gas is one of the products of the respiration of animals, so that the animal kingdom is one of the sources of plant carbon.

The transition from carbon dioxide to woody fibre is brought about in the plant by a series of chemical processes, and through the formation of a number of intermediate products in a manner which is not yet thoroughly understood; but since carbon dioxide consists of carbon and oxygen, and since plants feed upon carbon dioxide, appropriating the carbon and giving off the oxygen as a waste product, it is certain that work of some kind must be performed. This is evident, because it has been explained that when carbon combines with oxygen a great deal of heat is given out, and as this heat is the equivalent of the energy stored in the carbon, it follows from the doctrine of the Conservation of Energy, that in order to separate the carbon from the oxygen again, just the same amount of energy must be supplied as is evolved during the combustion of the carbon. If a pound of carbon in burning to carbon dioxide gives out heat equivalent to eleven million foot-pounds of work, we must apply the same amount of work to the carbon dioxide produced to separate it into its constituents. Neither a plant nor any living thing can create energy any more than it can create matter, and just as the matter composing a living organism is assimilated from external sources, so must we look to an external source for the energy which enables the plant to do this large amount of chemical work.

The separation of carbon from oxygen in the plant is effected by means of energy supplied by the sun. The great white hot globe which is the centre of our system, and round which this earth and the planets are moving, is a reservoir from which there is constantly pouring forth into space a prodigious quantity of energy. It must be remembered that the sun is more than a million times greater in bulk than our earth. It has been calculated by Sir William Thomson that every square foot of the sun's surface is radiating energy equivalent to 7000 horse-power in work. On a clear summer day the earth receives from the sun in our latitude energy equal to about 1450 horse-power per acre. To keep up this supply by the combustion of coal, we should have to burn for every square foot of the sun's surface between three and four pounds per second. A small fraction of this solar energy reaches our earth in the form of radiant heat and light, and it is the latter which enables the plant to perform the work of separating the carbon from the oxygen with which it is chemically combined. It is, in fact, well known that the growth of plants—that is, the assimilation of carbon and the libération of oxygen —only takes place under the influence of light. This function is performed by the leaves which contain the green colouring-matter known as chlorophyll, the presence of which is essential to the course of the chemical changes.

If we now sum up the results to which we have been led, it will be seen—

(1) That the chief source of the energy contained in coal is the carbon.

(2) That this carbon formed part of the plants which grew during the Carboniferous period.

(3) That the carbon thus accumulated was supplied to the plants by the carbon dioxide existing in the atmosphere at that time.

(4) That the separation of the carbon from the oxygen was effected in the presence of chlorophyll, by means of the solar energy transmitted to the earth during the Carboniferous period.

We thus arrive at the interesting conclusion, that

the heat which we get from coal is sunlight in another form. For every pound of coal that we now burn, and for every unit of heat or work that we get from it, an equivalent quantity of sunlight was converted into the latent energy of chemical separation during the time that the coal plant grew. This energy has remained stored up in the earth ever since, and reappears in the form of heat when we cause the coal to undergo combustion. It is related that George Stephenson when asked what force drove his locomotive, replied that it was "bottled-up sunshine," and we now see that he was much nearer the truth in making this answer than he could have been aware of at the time.

Before passing on to the consideration of the different products which we get from coal, it will be desirable to discuss a little more fully the nature of the change which occurs during the transformation of wood into coal. Pure woody fibre consists of a substance known to chemists as cellulose, which contains fifty per cent. of carbon, the remainder of the compound being made up of hydrogen and oxygen. It is thus obvious that during the fossilization of the wood some of the other constituents are lost, and the percentage of carbon by this means raised. We can trace this change from wood, through peat, lignite, and the different varieties of coal up to graphite, which is nearly pure carbon. It is in fact possible to construct a series showing the conversion of wood into coal, this series comprising the varieties given in the table on p. 23, as well as younger and older vegetable deposits. The series will be—

I. Woody fibre (cellulose).
II. Peat from Dartmoor.
III. Lignite, or brown coal, an imperfectly carbonized vegetable deposit of more recent geological age than true coal.
IV. Average bituminous coal.
V. Cannel coal from Wigan.
VI. Anthracite from Wales.
VII. Graphite, the oldest carbonaceous mineral.

The percentage of the chief elements in the members of this series is—

	Carbon.	Hydrogen.	Oxygen.
I.	50·0	6·0	44·0
II.	54·0	5·2	28·2
III.	66·3	5·6	22·8
IV.	77·0	5·0	11·2
V.	81·2	5·6	7·9
VI.	90·1	3·2	2·5
VII.	94—99·5, the remainder being ash.		

In the above table the increase of carbon and the decrease of oxygen is well brought out; the hydrogen also on the whole decreases, although with some irregularity. The exact course of the chemical change which occurs during the passage of wood into coal is at present involved in obscurity. The oxygen may be eliminated in the form of water or of carbon dioxide or both; some of the carbon is got rid of in the form of marsh gas, a compound of carbon and hydrogen, which forms the chief constituent of the dangerous "fire-damp" of coal mines.

Marsh gas is an inflammable gas which becomes explosive when mixed with air and ignited; it often escapes with great violence during the working of coal seams, the jets blowing out from the coal or underclay with a rushing noise, indicative of the high pressure under which the hydrocarbon gas has accumulated. These jets of escaping gas are known amongst miners as "blowers." If the air of a mine contains a sufficient quantity of the gas, and a flame accidentally fires the mixture, there results one of those disastrous explosions with which the history of coal mining has unfortunately only made us too familiar.

From the account of coal which has thus far been rendered, it will be seen that as a source of mechanical power, we are far from using it as economically as could be desired; and when we look at our open grates with clouds of unburnt carbon particles escaping up the chimney, and so constructed that only a small fraction of the total heat warms our rooms, it will be seen that the tale of waste is still more deplorable. But we are at present rather concerned with what we actually do get from coal than with what we ought to get from it, and here, when we come to deal with the various material products, we shall have a better account to present.

If instead of heating coal in contact with air and allowing it to burn, we heat it in a closed vessel, such as a retort, it undergoes decomposition with the formation of various gaseous, liquid, and solid products. This process of heating an organic compound in a closed vessel without access of air and collecting the products, is called destructive distillation. The tobacco-pipe experiment of our boyhood is our first practical introduction to the destructive distillation of coal. We put some powdered coal into the bowl of the pipe, plaster up the opening with clay and then insert the bowl in a fire, allowing the stem to project from between the bars of the grate. In a few minutes a stream of gas issues from the orifice of the stem; on applying a light it burns with a luminous flame, and we have made coal-gas on a small scale.

In the destructive distillation of organic substances, such as wood or coal, there are always produced four things — gas, watery liquid, and viscous products known as tar, while a residue of coke or charcoal is left in the retort. This is a very old observation, and was made so long ago that it becomes interesting as a point in the history of applied science to know who first submitted coal to destructive distillation. According to Dr. Gustav Schultz, we must credit a German with this observation, which was made towards the end of the seventeenth century (about 1680) by a chemist named Johann Joachim Becher. The experiment is described in such a quaint manner that the exact words of the author are worthy of being reproduced, and the passage is here given as translated by Dr. Lunge in his work on *Coal Tar and Ammonia*—

"In Holland they have peat, and in England pit-coals; neither of them is very good for burning, be it in rooms or for smelting. But I have found a way, not merely to burn both kinds into good coal (coke) which not any more smokes nor stinks, but with their flame to smelt equally well as with wood, so that a foot of such coal makes flames 10 feet long. That I have demonstrated with pit-coal at the Hague, and here in England at Mr. Boyles', also at Windsor on the large scale. In this connection it is also noteworthy that, equally as the Swedes make their tar from firwood, I have here in England made from pit-coal a sort of tar which is equal to the Swedish in every way, and for some operations is even superior to it. I have made proof of it on wood and on ropes, and the proof has been found right, so that even the king has seen a specimen of it, which is a great thing in England, and the coal from which the tar has been taken out is better for use than before."

This enterprising chemist, moreover, brought his results to a practical issue, for he secured a patent, in conjunction with Henry Serle, in 1681, for "a new way of making pitch and tarre out of pitcoale, never before found out or used by any other."

No less interesting is the work of our own clergy during the last century, when many eminent divines appear to have devoted their leisure to experimental science. Thus, about the year 1688 the Rev. John Clayton, D.D., Dean of Kildare, went to examine a ditch two miles from Wigan in Lancashire, the water in which had been stated to "burn like brandy" when a flame was applied to it. The Dean ultimately traced the phenomenon to an escape of inflammable gas from an underlying coal seam, and he followed up the matter experimentally by studying the destructive distillation of Wigan coal in retorts. The results were communicated to the Hon. Robert Boyle, but were not published till after the death of the latter, and long after the death of the author. The following account is taken from the abridged edition of the *Philosophical Transactions* (1739) :—

"At first there came over only phlegm, afterwards a black oil, and then also a spirit arose, which he could noways condense, but it forced the luting, or broke the glasses. Once, when it had forced the lute, coming close to it to try to repair it, he observed that the spirit which issued out caught fire at the flame of the candle, and continued burning with violence as it issued out in a stream, which he blew out and lighted again alternately for several times. He then tried to save some of this spirit. Taking a turbinated receiver, and putting a candle to the pipe of the receiver while the spirit rose, he observed that it caught flame, and continued burning at the end of the pipe, though you could not discern what fed the flame. He then blew it out, and lighted it again several times; after which he fixed a bladder, flatted and void of air, to the pipe of the receiver. The oil and phlegm descended into the receiver, but the spirit, still ascending, blew up the bladder. He then filled a good many bladders with it, and might have filled an inconceivable number more; for the spirit continued to rise for several hours, and filled the bladders almost as fast as a man could have blown them with his mouth ; and yet the quantity of coals he distilled was inconsiderable.

"He kept this spirit in the bladders a considerable time, and endeavoured several ways to condense it, but in vain. And when he wished to amuse his friends, he would take one of the bladders, and pricking a hole with a pin, and compressing gently the bladder near the flame of a candle till it once took fire, it would then continue flaming till all the spirit was compressed out of the bladder."[1]

The Rev. Stephen Hales, D.D., Rector of Farringdon, Hants, was the author of a book entitled *Statical Essays, containing Vegetable Staticks*, printed in 1726-27, and of which the third edition bears the date 1738. At p. 182 of this work, after a previous description of the destructive distillation of all kinds of substances in iron or other retorts, he says—

"By the same means also I found plenty of air [gas] might be obtained from minerals. Half a cubick inch, or 158 grains of Newcastle coal, yielded in distillation 180 cubick inches of air [gas], which arose very fast from the coal, especially while the yellowish fumes ascended."

Still later, viz. about 1767, we have the Rev. R. Watson, D.D., Regius Professor of Divinity in the University of Cambridge, and Bishop of Llandaff, interesting himself in chemistry. He wrote a series

[1] "An experiment concerning the Spirit of Coals," *Phil. Trans.* (abridged), vol. viii. p. 295.

of *Chemical Essays*, one of which is entitled, *Of Pit Coal*, and in this he describes the production from coal (by destructive distillation) of illuminating gas, ammonia-water, tar, and coke. He further compares the relative quantities of the different products from various kinds of coal, but he appears to have been chiefly interested in the tar, and disregarded the gas and other products. Not the least interesting part of his book is the preface, in which he apologizes for his pursuits in the following words—

"Divines, I hope, will forgive me if I have stolen a few hours, not, I trust, from the duties of my office, but certainly from the studies of my profession, and employed them in the cultivation of natural philosophy. I could plead in my defence, the example of some of the greatest characters that ever adorned either this University or the Church of England."

This is quoted from the 5th edition, dated 1789, the essay on coal being in the second of five volumes. As the learned bishop published other works on chemistry, we may suppose that the forgiveness which he asks from his brother divines was duly accorded.

None of these preliminary experiments, however, led to any immediate practical result so far as concerns the use of coal-gas as an illuminating agent. Towards the end of the last century the lighting of individual establishments commenced, and the way was thus prepared for the manufacture of the gas on a large scale. One of the earliest pioneers was the ninth Earl of Dundonald, an inventive genius, who in 1782 at Culross Abbey became one of the first practical tar distillers. He secured letters patent in 1781 for making tar, pitch, essential oils, volatile alkali, mineral acids, salts, and cinders from coal. The gas was only a waste product, and, strange as it may appear, the Earl, whose operations were financial failures, did not realize the importance of the gas, the tar and coke being considered the only products of value. Here is the account of the experiments by his son, Admiral Dundonald, the Sailor Earl, quoted from his *Autobiography of a Seaman* :—

"In prosecution of his coal-tar patent, my father went to reside at the family estate of Culross Abbey, the better to superintend the works on his own collieries, as well as others on the adjoining estates of Valleyfield and Kincardine. In addition to these works, an experimental tar-kiln was erected near the Abbey, and here coal-gas became accidentally employed in illumination. Having noticed the inflammable nature of a vapour arising during the distillation of tar, the Earl, by way of experiment, fitted a gun-barrel to the eduction pipe leading from the condenser. On applying fire to the muzzle, a vivid light blazed forth across the waters of the Frith, becoming, as was afterwards ascertained, distinctly visible on the opposite shore."

A few years later the foundation of the coal-gas manufacture was laid by William Murdoch, a Scotchman, who must be credited with the practical introduction of this illuminating agent. The idea had about the same time occurred to a Frenchman, Lebon, but in his hands the suggestion did not take a practical form. Murdoch was overseer of some mines in Cornwall, and in 1792 he first lighted his own house at Redruth. He then transferred his services to the great engineering firm of Boulton and Watt at Soho, near Birmingham, where he erected apparatus in 1798, and in the course of a few years the whole of this factory was permanently lighted by gas. From this time the introduction of gas into other factories at Manchester and Halifax was effected by Murdoch and his pupil, Samuel Clegg. From single factories coal-gas at length came into use as a street illuminant, although somewhat tardily. Experiments were made in London at the Lyceum Theatre in 1803, in Golden Lane in 1807, and in Pall Mall two years later.

It was fifteen years from the time of Murdoch's first installation at Soho before the streets of London were lighted by gas on a commercial scale. Our grandfathers seem to have had a great dread of gas, and public opposition no doubt had much to do with its exclusion from the metropolis. There were even at that time eminent literary and scientific men who did not hesitate to cast ridicule upon the proposal, and to declare the scheme to be only visionary. But about 1806 there came into this country an energetic German who passed by the name of Winsor, and who is described as an ignorant adventurer, whose real name was Winzler. Whatever his origin, he certainly helped to rouse the public interest in gas lighting. He took out a patent, he gave public lectures, and collected large sums of money for the establishment of gas companies. Most of the capital was, however, squandered in futile experiments, but at length in 1813, Westminster Bridge, and a year later St. Margaret's parish, was successfully lighted. From that period the use of gas extended, but it was some time before the public fears were allayed, for it is related that Samuel Clegg, who undertook the lighting of London Bridge, had at first to light his own lamps, as nobody could be found to undertake this perilous office. Even after gas had come into general use as a street illuminant, it must have found its way but slowly into private houses. In an old play-bill of the Haymarket Theatre, dated 1843—thirty years after the first introduction into the streets—it is announced—

"Among the most important Improvements, is the introduction (for the first time) of Gas as the Medium of Light!"

The manufacture of coal-gas, first rendered practicable by the energy and skill of the Scotch engineer Murdoch, is now carried on all over the country on a colossal scale. It is not the province of the present volume to deal with the details of manufacture, but a short description of the process is necessary for the proper understanding of the subsequent portions of the subject (see Fig. 3). The coal is heated in clay cylinders, called retorts, provided with upright exit pipes through which the volatile products escape, and are conducted into water contained in a horizontal pipe termed the "hydraulic main." In the latter the gas is partially cooled, and deposits most of the tar and watery liquor which distil over at the high temperature to which the retorts are heated. The tar and watery liquor are allowed to flow from the hydraulic main into a pit called the "tar well," and the gas then passes through a series of curved pipes exposed to the air, in which it is further cooled, and deposits more of the tar. From this "atmospheric condenser" the gas passes into a series of vessels filled with coke, down which a fine spray of water is constantly being blown. These vessels, known as "scrubbers," serve to remove the last traces of tar, and some of

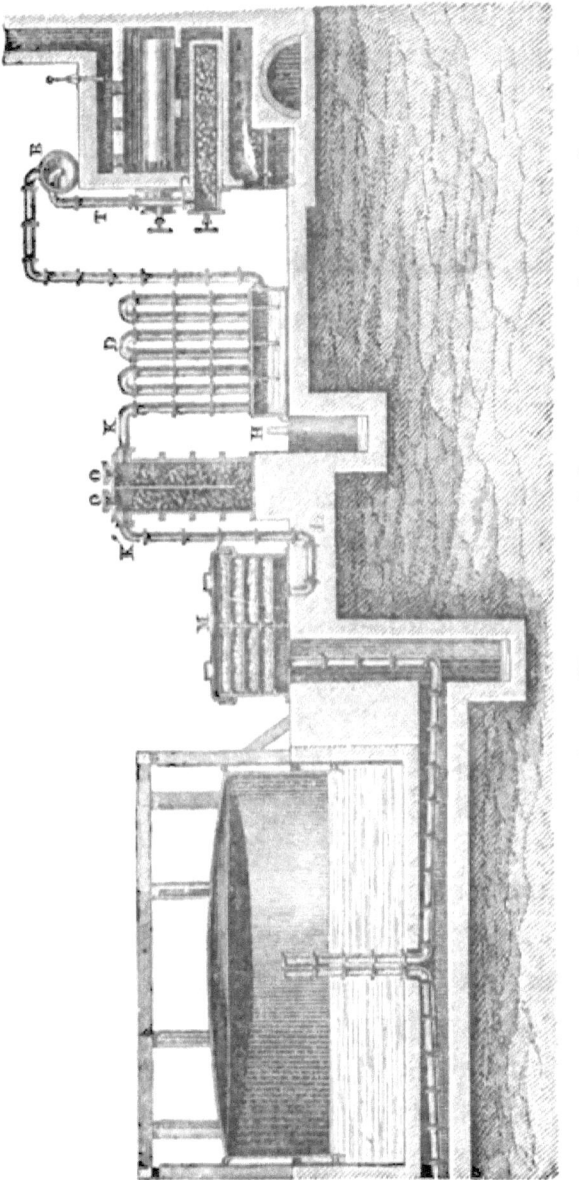

FIG. 3.—Sectional diagram of gas plant. The retorts and furnace are on the right; the gas rises through the upright pipe T into the hydraulic main B; from there it passes into the atmospheric condensers D, from the lower cistern of which the condensed tar flows into the tar-well, H. Passing up through K, the gas is conducted into the scrubber, O, and from there into the purifier, M. From there it emerges through K′ into the purifier, M, and then into the gas-holder for distribution. (From Schultz's *Chemie des Steinkohlentheers*.)

the volatile sulphur compounds which are formed from the small quantity of sulphur present in most coals. The removal of sulphur compounds is a matter of importance, because when gas is burnt these compounds give rise to acid vapours, which are deleterious to health and destructive to property.

From the scrubbers the gas is sent through another series of vessels packed with trays of lime or oxide of iron, in order to remove sulphuretted hydrogen and other sulphur compounds as completely as possible. A small quantity of carbon dioxide is also removed by these "purifiers," as the presence of this gas impairs the illuminating power of coal-gas. From the purifiers the gas passes into the gas-holders, where it is stored for distribution. It remains only to be stated that the distillation of coal is effected under a pressure somewhat less than that of the atmosphere, the products of distillation being pumped out of the retorts by means of a kind of air-pump, called an exhauster, which is interpolated between the hydraulic main and the condenser, or at some other part of the purifying system. The coke left in the retort is used as fuel for burning under the retorts or for other purposes. The oxide of iron used in the purifiers can be used over and over again for a certain number of times by exposing it to the air, and when it is finally exhausted, the sulphur can be burnt out of it and used for making that most important of all chemical products, sulphuric acid. Thus the small quantity of sulphur present in the original coal (probably in the form of iron pyrites) is rendered available for the manufacture of a useful product.

The necessarily brief description of this important industry will suffice for the general reader. Those who desire further information on points of detail will refer to special works. We are here rather concerned with the subsequent fate of the different products, four of which have to be dealt with, viz. the gas, watery liquor, tar, and coke. The first and last of these having already been accounted for—the one as an illuminating agent and the other as fuel—may now be dismissed.

No story of applied science is complete unless we can form some idea of the quantities of material used, and the amount of the products obtained. From one ton of Newcastle coal we get about 10,000 cubic feet of gas, 110 to 120 lbs. of tar, 20 to 25 gallons of watery liquor, and about 1500 lbs. of coke. Different coals of course give different quantities, and the latter vary also according to the heat of distillation; but the above estimate will furnish a good basis for forming our ideas with some approach to precision. It has been estimated also that we are now distilling coal at the rate of about ten million tons per annum, so that there is annually produced 100,000 million cubic feet of gas, and about 500,000 tons of tar, besides proportionate quantities of the other products. The great metropolitan companies alone are consuming nearly three million tons annually for the production of gas, a consumption corresponding to about 6000 cubic feet per head of the population. This of course takes no account of the coal used for other manufactures or for domestic purposes, but it is interesting to compare these estimates with the consumption of coal in London about a century ago, before the introduction of gas, when, as Bishop Watson tells us in his work already referred to, the annual consumption was 922,394 tons.

The enormous quantity of tar resulting from our gas manufacture furnishes the raw material for the production of a multitude of valuable substances—colouring-matters, medicines, perfumes, flavouring-matters, burning and lubricating oils, &c. Out of this unsavoury waste material of the gas-works, the researches of chemists have enabled a great industry to spring up which is of continually growing importance. It will be the object of the remaining portion of the present volume to set forth the achievements of science in this branch of its application. The foundation of the coal-tar industry was laid in this country—the country where coal was first distilled on a large scale for the production of gas, and where the first of the coal-tar colouring-matters was sent forth into commerce. We are at the present time the largest tar producers in Europe; it has been stated that we produce more than double the whole quantity of tar made in the gas-using countries of Europe; but in spite of this, our manufacture of finished products is by no means in that flourishing condition which might be expected from our natural resources in the way of coal, and the facilities which we possess for manufacturing the raw materials out of it.

But we must now take a glance at some of the other uses to which coal is put in order to realize more completely the truth of the statement made some pages back, viz. that this mineral has been the chief source of our industrial prosperity. Great as is the consumption of coal by the gas manufacturer, there is an equal or even a greater demand for the carbonaceous residue left when the coal has been decomposed by destructive distillation or by partial combustion. This residue is coke—the substance left in the retorts after the gas manufacture. There is a great demand for coke for many purposes; it is used in most cases where a cheap smokeless fuel is required; it is burnt in the furnaces of locomotives and other engines, and is very largely consumed by the iron smelter in the blast furnace.

To meet these demands a large quantity of coal is converted into coke by being burnt in ovens with an insufficient supply of air for complete combustion, or in suitably constructed close furnaces. The tar and other products have in this country until recently been allowed to escape as waste, but the time is approaching when these must be utilized. It will give an idea of the industrial importance of coke when it is stated, that about twelve million tons of our coal annually undergo conversion into this form of fuel. Chemically considered, coke consists of carbon together with all the mineral constituents of the coal, and small quantities of hydrogen, oxygen, and nitrogen. The amount of carbon varies from 85 to 97, and the ash from 3 to 14 per cent.

The conversion of coal into coke is a very venerable branch of manufacture, which was first carried out on a large scale in this country about the middle of the seventeenth century. As an operation it may appear utterly devoid of romance, but as Goethe has described his visit to the earliest of coke-burners, this fragment of history is worth narrating. When the great German philosophical poet was a student at Strassburg (1771), he rode over with some friends to visit the neighbourhood of Saarbrücken where he met an old "coal philosopher" named Stauf, who was there carrying on the industry. This "philosophus per ignem" was manager of some alum works, and the ruling spirit of the "burning hill" of Duttweiler. The hill no doubt owed its designation to the coke ovens at work upon it, and which had been in operation there for some six or seven years before Goethe's visit, *i.e.* since 1764. The coke was wanted for iron smelting, and even at that early period Stauf had the wisdom to condense his volatile products, for we are told that he showed his visitors bitumen, burning-oil, lampblack, and even a cake of sal ammoniac resulting from his operations. Goethe has put upon record his visit to the little haggard old coke-burner, living in his lonely cottage in the forest (*Aus meinem Leben: Wahrheit und Dichtung,*" Book X). It is probably Stauf's ovens which are described by the French metallurgist, De Gensanne, in his *Traité de la fonte des Mines par le feu du Charbon de Terre*, published in Paris in 1770. After long years of coke-making, without any regard to the value of the volatile products, we are now beginning to consider the advisability of doing that which has long been done on the Continent.

It is not unlikely that Bishop Watson in the

last century had heard of the attempt to recover the products from coke ovens, for he gives the following very sound advice in his *Chemical Essays* :—

"Those who are interested in the preparation of coak would do well to remember that every 96 ounces of coal would furnish four ounces at the least of oil, probably six ounces might be obtained; but if we put the product so low as five ounces from 100, and suppose a coak oven to work off only 100 tons of coal in a year, there would be a saving of five tons of oil, which would yield above four tons of tar; the requisite alteration in the structure of the coak ovens, so as to make them a kind of distilling vessels, might be made at a very trifling expense."—5th ed., 1789, vol. ii. p. 351.

We have yet to chronicle another chapter in the history of coal philosophy before finishing with this part of the subject. There is a branch of manufacture carried on, especially in Scotland, which results in the production of burning and lubricating oils, and solid paraffin, a wax-like substance which is used for candle-making. The manufacture of candles out of coal will perhaps be a new revelation to many readers of this book. It must be admitted, however, that the term "coal" is here being extended to only partially fossilized vegetation of younger geological age than true coal, and to bituminous shales of various ages. Shale, geologically considered, is hardened mud; it may be looked upon as clay altered by time and pressure. Now if the mud, at the period of its deposition, was much mixed up with vegetable matter, we should have in course of time a mixture of more or less carbonized woody fibre with mineral matter, and this would be called a carbonaceous or bituminous shale. Shales of this kind often contain as much as 80 to 90 per cent. of mineral matter, and seldom more than 20 per cent. of volatile matter, *i.e.* the portion lost on ignition, and consisting chiefly of the carbonaceous constituents.

The story of the shale-oil industry is soon told. About the year 1847 oil was "struck" in a coal mine at Alfreton in Derbyshire, and in the hands of Mr. James Young this supply furnished the market with burning-oil for nearly three years. Then the spring became exhausted, and Mr. Young and his associates had to look out for another source of oil. Be it remembered that this happened some nine years before the utilization of the great American petroleum deposits. Many kinds of vegetable matter were submitted to destructive distillation before a substance was found which could be profitably worked, but at length Mr. Young tried a kind of cannel coal which had about that time been introduced for gas making. This substance was called Boghead gas coal or Torbane Hill mineral, from the place where it occurred, which is at Bathgate in Linlithgow. This mineral was found to yield a large amount of paraffin oil and solid paraffin on destructive distillation, and from that time (1850) to this, the industry has been carried on at Bathgate and other parts of Scotland, where similar carbonaceous deposits occur.

It may seem a matter of unimportance at the present time whether this Torbane Hill mineral is a true coal or not. About forty years ago, however, the decision of this question involved a costly law-suit in Edinburgh. The proprietor of the estate had granted a lease to a firm, conveying to the latter the right to work coal, limestone, ironstone, and certain other minerals found thereon, but excluding copper and all other minerals not mentioned in the contract. The lessees then found that this particular carbonaceous mineral was of very great value, both on account of the high quality of the gas, and afterwards on account of the paraffin which it furnished by Young's process of distillation. Thereupon the lessor brought an action against the lessees, claiming £10,000 damages, on the ground that the latter had broken the contract by removing a mineral which was not coal. Experts gave evidence on both sides; some declared in favour of the substance being coal, others said it was a bituminous shale, while others called it bituminated clay, or refused to give it a name at all. Judgment was finally given for the defendants, so that in the eye of the law the mineral was considered a true coal. As a matter of fact, it is impossible to draw a hard and fast line between coal and bituminous shale, as the one is connected with the other by a series of intermediate minerals, and the Torbane Hill mineral happens to form one of the links. It contains about 69 per cent. of volatile matter, and leaves 31 per cent. of residue, consisting of 12 parts of carbon and 19 of ash.

The manufacture started by Young has developed into an important industry, in spite of the fact that the original Torbane Hill coal has become exhausted, and that enormous natural deposits of petroleum are worked in America, Russia, and elsewhere. There are now some fifteen companies at work in Scotland, representing an aggregate capital of about two and a half million pounds sterling. Bituminous shales of different kinds are distilled at a low red heat in iron retorts, and from the volatile portions there are separated those valuable products which have already been alluded to, viz. burning and lubricating oils, solvent mineral oil, paraffin wax for candles, and ammonia. We may fairly claim these as coal products, although the shales used contain much mineral matter, the carbon averaging about 20 per cent., the hydrogen three per cent., the nitrogen 0·7, and the ash about 67 per cent. The shales worked are approximately of the same age as true coal, *i.e.* Carboniferous. The Scotch companies are distilling about two million tons of shale per annum, this quantity producing about sixty million gallons of crude oil, and giving employment to over 10,000 hands.

It is not the province of the present work to enter into the chemical nature of the products of destructive distillation in any greater detail than is necessary to enable the general reader to know something of the recent discoveries in the utilization of these products. We shall, however, have occasion later on to make ourselves acquainted with the names of some of the more important raw materials which are derived from this source, and certain preliminary explanations are indispensable. In the first place then, let us start from the fact that coal—including carbonaceous shale and lignite—when heated in a closed vessel gives gas, tar, coke, and a watery liquor. A clear understanding must be arrived at concerning the manner in which these products arise.

There is a widely-spread notion that the substances derived from coal and utilized for industrial purposes are present in the mineral itself, and that the art of the chemist has been exercised in separating the said substances by various processes. This idea must be at once dispelled. It is true that there is a small quantity of water and a certain amount of gas already present in most coals, but these are quite insignificant as compared with the total yield of gas and watery liquor. So also with respect to the tar; it is possible that in some highly bituminous minerals we might dissolve out a small quantity of tarry matter by the use of appropriate solvents, but in the coals mostly used for gas-making not a trace of tar exists ready formed, and still less can it be said that the coal contains coke. All these products are formed *by the chemical decomposition of the coal* under the influence of heat, and their nature and quantity can be made to vary within certain limits by modifying the temperature of distillation.

Having once realized this principle with respect to coal itself, it is easy to extend it to the products of its destructive distillation. The tar, for instance, is a complicated mixture of various substances, among which hydrocarbons—*i.e.* compounds of carbon with hydrogen—largely predominate. The different components of coal-tar can be separated by processes which we shall have to consider subsequently. Of the compounds thus isolated some few are immediately applicable for industrial purposes, but the majority only form the raw materials for the manufacture of other products, such as colouring-matters and medicines. Now these colouring-matters and other finished products no more exist in the tar than the latter exists in the coal. They are produced from the hydrocarbons, &c., present in the tar *by chemical processes*, and bear much about the same relationship to their parent substances that a steam-engine bears to the iron ore out of which its metallic parts are primarily constructed. Just as the mechanical skill of the engineer enables him to construct an engine out of the raw material iron, which is extracted from its ore, and converted into steel by chemical processes, so the skill of the chemist enables him to build up complex colouring-matters, &c., out of the raw materials furnished by tar, which is obtained from coal by chemical decomposition.

The illuminating gas which is obtained from coal by destructive distillation consists chiefly of hydrogen and gaseous hydrocarbons, the most abundant of the latter being marsh gas. There are also present in smaller quantities the two oxides of carbon, the monoxide and the dioxide, which are gaseous at ordinary temperatures, together with other impurities. Coal-gas is burnt just as it is delivered from the mains—it is not at present utilized as a source of raw material in the sense that the tar is thus made use of. In some cases gas is used as fuel, as in gas-stoves and gas-engines, and in the so-called "gas-producers," in which the coal, instead of being used as a direct source of heat, is partially burnt in suitable furnaces, and the combustible gas thus arising, consisting chiefly of carbon monoxide, is conveyed to the place where it undergoes complete combustion, and is thus utilized as a source of heat.

Summing up the uses of coal thus far considered, we see that this mineral is being consumed as fuel, for the production of coke, for the manufacture of gas, and in many other ways. Lavishly as Nature has provided us with this source of power and wealth, the idea naturally suggests itself whether we are not drawing too liberally upon our capital. The question of coal supply crops up from time to time, and the public mind is periodically agitated about the prospects of its continuance. How long we have been draining our coal resources it is difficult to ascertain. There is some evidence that coal-mining was carried on during the Roman occupation. In the reign of Richard I. there is distinct evidence of coal having been dug in the diocese of Durham. The oldest charters take us back to the early part of the thirteenth century for Scotland, and to the year 1239 for England, when King Henry III. granted a right of sale to the townsmen of Newcastle. With respect to the metropolis, Bishop Watson, on the authority of Anderson's *History of Commerce*, states that coal was introduced as fuel at the beginning of the fourteenth century. In these early days, when it was brought from the north by ships, it was known as "sea-coal":—

"Go; and we'll have a posset for 't soon at night, in faith, at the latter end of a sea-coal fire."—*Merry Wives of Windsor*, Act I., Sc. iv.

That the fuel was received at first with disfavour appears from the fact that in the reign of Edward I. the nobility and gentry made a complaint to the king objecting to its use, on the ground of its being a public nuisance. By the middle of the seventeenth century the use of coal was becoming more general in London, chiefly owing to the scarcity of wood; and its effects upon the atmosphere of the town will be inferred from a proclamation issued in the reign of Elizabeth, prohibiting its use during the sitting of Parliament, for fear of injuring the health of the knights of the shire. About 1649 the citizens again petitioned Parliament against the use of this fuel on account of the stench; and about the beginning of that century "the nice dames of London would not come into any house or roome when sea-coales were burned, nor willingly eat of meat that was either sod or roasted with sea-coale fire" (*Stow's Annals*).

For many centuries therefore we have been drawing upon our coal supplies, and using up the mineral at an increasing rate. According to a recent estimate by Professor Hull, from the beginning of the present century to 1875 the output has been more than doubled for each successive quarter century. The actual amount of coal raised in the United Kingdom between 1882 and the present time averages annually about 170 million tons, corresponding in money value to about £45,000,000 per annum. In 1860 the amount of coal raised in Great Britain was a little more than 80 million tons, and Professor Hull estimated that at that rate of consumption our supplies of workable coal would hold out for a thousand years. Since then the available stock has been diminished by some 3,650 million tons, and even this deduction, we are told on the same authority, has not materially affected our total supply. The possibility of a coal famine need, therefore, cause no immediate anxiety; but we cannot "eat our loaf and have it too," and sooner or later the continuous drain upon our coal resources must make itself felt. The first effect will probably be an increase in price owing to the greater depth at which the coal will have to be worked.

The whole question of our coal supply has, however, recently assumed a new aspect by the discovery (February 1890) of coal at a depth of 1,160 feet at Dover. To quote the words of Mr. W. Whitaker—"It may be indeed that the coal supply of the future will be largely derived from the South-East of England, and some day it may happen, from the exhaustion of our northern coal-fields, that we in the south may be able successfully to perform a task now proverbially unprofitable—*we may carry coal to Newcastle.*"

The coal-fields of Great Britain and Ireland occupy, in round numbers, an area of 11,860 square miles, or about one-tenth of the whole area of the land surface of the country. Within this area, and down to a depth of 4,000 feet, lie the main deposits of our available wealth. Some idea of the amount of coal underlying this area will be gathered from the table [1] on the next page.

This supply, amounting to over 90,000 million tons, refers to the exposed coal-fields and to workable seams, *i.e.* those above one foot in thickness. But in addition to this, we have a large amount of coal at workable depths under formations of later geological age than the Carboniferous, such as the Permian formation of northern and central England. Adding the estimated quantity of coal

[1] *Report of the Coal Commissioners* (1866-71), vol. i.

Coal Fields of—	Amount of coal in millions of tons to depths not exceeding 4,000 feet.
South Wales	32,456
Forest of Dean	265
Bristol	4,219
Warwickshire	459
S. Staffordshire, Shropshire, Forest of Wyre and Clee Hills	1,906
Leicestershire	837
North Wales	2,005
Anglesey	5
N. Staffordshire	3,825
Lancashire and Cheshire	5,546
Yorkshire, Derbyshire, and Northumberland	18,172
Black Burton	71
Northumberland and Durham	10,037
Cumberland	405
Scotland	9,844
Ireland	156

from this source to that contained in the exposed coal-fields as given above, we arrive at the total available supply. This is estimated to be about 146,454 million tons. To this we may one day have to add the coal under the south-eastern part of England.

It is important to bear in mind, that out of the 170 million tons of coal now being raised annually we only use a small proportion, viz. from 5 to 6 per cent. for gas-making. The largest amount (33 per cent.) is used for iron-smelting,[1] and about 15 per cent. is exported; the remainder is consumed in factories, dwelling-houses, for locomotion, and in the smaller industries.

The enormous advancement which has taken place of late years in the industrial applications of electricity has given rise to the belief that coal-gas will in time become superseded as an illuminating agent, and that the supply of tar may in consequence fall off. So far, however, the introduction of electric-lighting has had no appreciable effect upon the consumption of gas, and even when the time of general electric-lighting arrives there will arise as a consequence an increased demand for gas as a fuel in gas engines. Moreover, the use of gas for heating and cooking purposes is likely to go on increasing. Nor must it be forgotten that the quantity of tar produced in gas-works is now greater than is actually required by the colour-manufacturer, and much of this by-product is burnt as fuel, so that if the manufacture of gas were to

---

[1] In a paper read before the Royal Statistical Society by Mr. Price-Williams in 1889, this author points out that, owing to the introduction of the Bessemer process and other economical improvements, the amount of coal used in the iron and steel manufacture had fallen in 1867 to about sixteen and a half per cent. of the total quantity raised.

suffer to any considerable extent there would still be tar enough to meet our requirements at the present rate of consumption of the tar-products. Then again, the value of the tar, coke, and ammoniacal liquor is of such a proportion as compared with the cost of the raw material, coal, that there is a good margin for lowering the price of gas when the competition between the latter and electricity actually comes about. It will not then be only a struggle between the two illuminants, but it will be a question of electricity *versus* gas, *plus* tar and ammonia.

While the electrician is pushing forward with rapid strides, the chemist is also moving onwards, and every year witnesses the discovery of new tar products, or the utilization of constituents which were formerly of little or no value. Thus if the cost of generating and distributing electricity is being lowered, on the other hand the value of coal tar is likely to go on advancing, and it would be rash to predict which will come out triumphant in the end. But even if electricity were to gain the day it would be worth while to distil coal at the pit's mouth for the sake of the by-products, and there is, moreover, the tar from the coke ovens to fall back upon—a source which even before the use of coal-gas the wise Bishop of Llandaff advised us not to neglect.

CHAPTER II.

The nature of the products obtained by the destructive distillation of coal varies according to the temperature of distillation, and the age or degree of carbonization of the coal. The watery liquor obtained by the dry distillation of wood is acid, and contains among other things acetic acid, which is sometimes prepared in this way, and from its origin is occasionally spoken of as "wood vinegar." The older the wood, the more complete its degree of conversion into coal, and the smaller the quantity of oxygen it contains, the more alkaline does the watery liquid become. Thus the gas-liquor is distinctly alkaline, and contains a considerable quantity of ammonia, besides other volatile bases. The uses of ammonia are manifold, and nearly our whole supply of this valuable substance is now derived from gas-liquor. The presence of ammonia in this liquor is accounted for when it is known that this compound is a gas composed of nitrogen and hydrogen. It has already been explained that coal contains from one to two per cent. of nitrogen, and during the process of distillation about one-fifth of this nitrogen is converted into ammonia, the remainder being converted partly into other bases, while a small quantity remains in the coke.

Ammonia, the "volatile alkali" of the old chemists, and its salts are of importance in pharmacy, but the chief use of this compound is to supply nitrogen for the growth of plants. Plants must have nitrogen in some form or another, and as they cannot assimilate it *directly* from the atmosphere where it exists in the free state, some suitable nitrogen compound must be supplied to the soil. It is possible that certain leguminous plants may derive their nitrogen from the atmosphere through the intervention of micro-organisms, which appear capable of fixing free nitrogen and of supplying it to the plant upon whose roots they flourish. But this is second-hand nitrogen so far as concerns the plant. It is true also that the atmosphere contains small traces of ammonia and acid oxides of nitrogen, which are dissolved by rain and snow, and thus get washed down into the soil. These are the natural sources of plant nitrogen. But in agricultural operations, where large crops have to be raised as rapidly as possible, some

E

additional source of nitrogen must be supplied, and this is the object of manuring the soil.

A manure, chemically considered, is a mixture of substances capable of supplying the necessary nitrogenous and mineral food for the nourishment of the growing plant. The ordinary farm or stable manure contains decomposing nitrogenous organic matter, in which the nitrogen is given off as ammonia, and thus furnishes the soil with which it is mixed with the necessary fertilizer. But the supply of this manure is limited, and we have to fall back upon gas-liquor and native nitrates to meet the existing wants of the agriculturist. Important as is ammonia for the growth of vegetation, it is not in this form that the majority of plants take up their nitrogen. Soluble nitrates are, in most cases, more efficient fertilizers than the salts of ammonia, and the ammonia which is supplied to the soil is converted into nitrates therein before the plant can assimilate the nitrogen. The oxidation of ammonia into nitric acid takes place by virtue of a process called "nitrification," and there is very good reason for believing that this transformation is the work of a micro-organism present in the soil. The gas liquor thus supplies food to a minute organism which converts the ammonia into a form available for the higher plants. Some branches of agriculture—such as the cultivation of the beet for sugar manufacture—are so largely dependent upon an artificial source of nitrogen, that their very existence is bound up with the supply of ammonia salts or other nitrogenous manures. The relationship between the manufacture of beet-sugar and the distillation of coal for the production of gas is thus closer than many readers will have imagined; for while the supply of native guano or nitrate is uncertain, and its freight costly on account of the distance from which it has to be shipped, the sulphate of ammonia from gas-liquor is always at hand, and available for the purposes of fertilization.

Then again, there are other products of industrial value which are associated with ammonia, such, for example, as ammonia-alum and caustic soda. This last is one of the most important chemical compounds manufactured on a large scale, and is consumed in enormous quantities for the manufacture of paper and soap, and other purposes. Salts of this alkali are also essential for glass making. Of late years a method for the production of caustic soda has been introduced which depends upon the use of ammonia, and as this process is proving a formidable rival to the older method of alkali manufacture, it may be said that such indispensable articles as paper, soap, and glass are now to some extent dependent upon gas-liquor, and may in course of time become still more intimately connected with the manufacture of coal-gas.

But quantitative statements must be given in order to bring home to general readers the actual value of the small percentage of nitrogen present in coal. Thus it has been estimated, that one ton of coal gives enough ammonia to furnish about 30 lbs. of the crude sulphate. The present value of this salt is roughly about £12 per ton. The ten million tons of coal distilled annually for gas making would thus give 133,929 tons of sulphate, equal in money value to £1,607,148, supposing the whole of the ammonia to be sold in this form. To this may be added the ammonia obtained during the distillation of shale and the carbonization of coal for coke, the former source furnishing about 22,000 tons, and the latter about 2500 tons annually. Small as is the legacy of nitrogen bequeathed to us from the Carboniferous period, we see that it sums up to a considerable annual addition to our industrial resources.

The three products resulting from the distillation of coal—viz. the gas, ammoniacal-liquor, and coke —having now been made to furnish their tale, we have next to deal with the tar. In the early days of gas manufacture this black, viscid, unsavoury substance was in every sense a waste product. No use had been found for it, and it was burnt, or otherwise disposed of. No demand for the tar existed which could enable the gas manufacturers to get rid of their ever-increasing accumulation. Wood-tar had previously been used as a cheap paint for wood and metal-work, and it was but a natural suggestion that coal-tar should be applied to the same purposes. It was found that the quality of the tar was improved by getting rid of the more volatile portions by boiling it in open pans; but this waste—to say nothing of the danger of fire—was checked by a suggestion made by Accum in 1815, who showed that by boiling down the tar in a still instead of in open pans the volatile portions could be condensed and collected, thus furnishing an oil which could be used by the varnish maker as a substitute for turpentine. A few years later, in 1822, the distillation of tar was carried on at Leith by Drs. Longstaff and Dalston, the "spirit" being used by Mackintosh of Glasgow for dissolving india-rubber for the preparation of that waterproof fabric which to this day bears the name of the original manufacturer. The residue in the still was burnt for lamp-black. Of such little value was the tar at this time that Dr. Longstaff tells us that the gas company gave them the tar on condition that they removed it at their own expense. It appears also that tar was distilled on a large scale near Manchester in 1834, the "spirit" being used for dissolving the residual pitch so as to make a black varnish.

But the production of gas went on increasing at a greater rate than the demand for tar for the above-mentioned purposes, and it was not till 1838 that a new branch of industry was inaugurated, which converted the distillation of this material from an insignificant into an important manufacture. In that year a patent was taken out by Bethell for preserving timber by impregnating it with the heavy oil from coal-tar. The use of tar for this purpose had been suggested by Lebon towards the end of the last century, and a patent had been granted in this country in 1836 to Franz Moll for this use of tar-products. But Bethell's process was put into a working form by the great improvements in the apparatus introduced by Bréant and Burt, and to the latter is due the credit of having founded an industry which is still carried on by Messrs. Burt, Bolton and Haywood on a colossal scale. The "pickling" or "creosoting" of timber is effected in an iron cylindrical boiler, into which the timber is run; the cylinder being then closed the air is pumped out, and the air contained in the pores of the wood thus escapes. The creosoting oil, slightly warmed, is then allowed to flow into the boiler, and thus penetrates into the pores of the wood, the complete saturation of which is insured by afterwards pumping air into the cylinder and leaving the timber in the oil for some hours under a pressure of 8 to 10 atmospheres.

All timber which is buried underground, or submerged in water, is impregnated with this antiseptic creosote in order to prevent decay. It will be evident that this application of tar-products must from the very commencement have had an enormous influence upon the distillation of tar as a branch of industry. Consider the miles of wooden sleepers over which our railways are laid, and the network of telegraph wires carried all over the country by wooden poles, of which the ends are buried in the earth. Consider also the many subaqueous works which necessitate the use of timber, and we shall gain an idea of the demand for heavy coal-tar oil created by the introduction of Bréant's process. Under the treatment described a cubic foot of wood absorbs about a gallon of oil, and by far the largest quantity of the tar oils is consumed in this way at the present time. Now in the early days of timber-pickling the lighter oils of the tar, which first come over on distillation, and which are too volatile for the purpose of creosoting, were in much about the same industrial position as the tar itself before its application as a timber preservative. The light oil had a limited use as a solvent for waterproofing and varnish making, and a certain quantity was burnt as coal-tar naphtha in specially constructed lamps, the invention of the late Read Holliday of Huddersfield, whose first patent was taken out in 1848 (see Fig. 4). Up to this time, be it remembered, that chemists had not found out what this naphtha contained. But science soon laid hands on the materials furnished by the tar-distiller, and the naphtha was one of the first products which was made to reveal the secret of its hidden treasures to the scientific investigator. From this period science and industry became indissolubly united, and the researches of chemists were

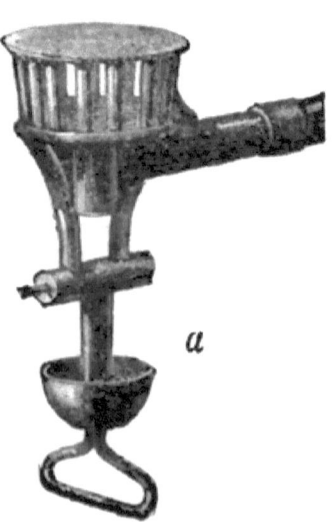

FIG. 4.—Read Holliday's lamp for burning light coal-tar oils. The oil is contained in the cistern c, from whence it flows down the pipe, when the stopcock is opened, into the burner a. Below the burner is a little cup, in which some of the oil is kept burning, and the heat from this flame volatilizes the oil as it flows down the pipe, the vapour thus generated issuing from the jets in the burner and there undergoing ignition. The burner and cup are shown on an enlarged scale at a in the lower figure.

carried on hand-in-hand with the technical developments of coal-tar products.

In 1825 Michael Faraday discovered a hydrocarbon in the oil produced by the condensation of "oil gas"—an illuminating gas obtained by the destructive distillation of oleaginous materials. This hydrocarbon was analysed by its illustrious discoverer, and named in accordance with his results "bicarburet of hydrogen." In 1834 the same hydrocarbon was obtained by Mitscherlich by heating benzoic acid with lime, and by Péligot by the dry distillation of calcium benzoate. For this reason the compound was named "benzin" by Mitscherlich, which name was changed into "benzol" by Liebig. In this country the hydrocarbon is known at the present time as benzene. Twenty years after Faraday's discovery, viz. in 1845, Hofmann proved the existence of benzene in the light oils from coal-tar, and in 1848 Hofmann's pupil, Mansfield, isolated considerable quantities of this hydrocarbon from the said light oils by fractional distillation. At the time of these investigations no great demand for benzene existed, but the work of Hofmann and Mansfield prepared the way for its manufacture on a large scale, when, a few years later, the first coal-tar colouring-matter was discovered by our countryman, W. H. Perkin.

It is always of interest to trace the influence

of scientific discovery upon different branches of industry. As soon as it had been shown that benzene could be obtained from coal-tar, the nitro-derivative of this hydrocarbon—*i. e.* the oily compound produced by the action of nitric acid upon benzene—was introduced as a substitute for bitter almond oil under the name of "essence of mirbane." Nitrobenzene has an odour resembling that of bitter almond oil, and it is still used for certain purposes where the latter can be replaced by its cheaper substitute, such as for the scenting of soap. Although the isolation of benzene from coal-tar gave an impetus to the manufacture of nitrobenzene, no use existed for the latter beyond its very limited application as "essence of mirbane," and the production of this compound was at that time too insignificant to take rank as an important branch of chemical industry.

The year 1856 marks an epoch in the history of the utilization of coal-tar products with which the name of Perkin will ever be associated. In the course of some experiments, having for their object the artificial production of quinine, this investigator was led to try the action of oxidizing agents upon a base known as aniline, and he thus obtained a violet colouring matter—the first dye from coal-tar —which was manufactured under a patent granted in 1858, and introduced into commerce under the name of mauve. A brief sketch of the history of aniline will serve to show how Perkin's discovery gave a new value to the light oils from coal-tar and raised the manufacture of nitrobenzene into an important branch of industry.

Thirty years before Perkin's experiments the Dutch chemist Unverdorben obtained (1826) a liquid base by the distillation of indigo, which had the property of forming beautifully crystalline salts, and which he named for this reason "crystallin." In 1834 Runge discovered the same base in coal-tar, although its identity was not known to him at the time, and because it gave a bluish colour when acted upon by bleaching-powder, he called it "kyanol." Again in 1840, by distilling a product obtained by the action of caustic alkalies upon indigo, Fritzsche prepared the same base, and gave it the name of aniline, from the Spanish designation of the indigo plant, "anil," derived from the native Indian word, by which name the base is known at the present time. That aniline could be obtained by the reduction of nitrobenzene was shown by Zinin in 1842, who used sulphide of ammonium for reducing the nitrobenzene, and named the resulting base "benzidam." The following year Hofmann showed that crystallin, kyanol, aniline, and benzidam were all one and the same base. Thus when the discovery of mauve opened up a demand for aniline on the large scale, the labours of chemists, from Unverdorben in 1826 to Hofmann in 1843, had prepared the way for the manufacturer. It must be understood that although Runge had discovered aniline in coal-tar, this is not the source of our present supply, for the quantity is too small to make it worth extracting. A mere trace of aniline is present in the tar ready formed; from the time this base was wanted in large quantities it had to be made by nitrating benzene, and then reducing the nitrobenzene.

The light oils of tar distillation rejected by the timber-pickling industry now came to the front, imbued with new interest to the technologist as a source of benzene for the manufacture of aniline. The inauguration of this manufacture, like the introduction of steam locomotion, is connected with a sad catastrophe. Mansfield, who first showed manufacturers how to separate benzene and other hydrocarbons from the light oils of coal-tar, and who devised for this purpose apparatus similar in principle to that used on a large scale at the present time (see Fig. 5), met with an accident which resulted in his death. In the upper part of a house in Holborn in February 1856, this pioneer was carrying on his experiments, when the contents of a still boiled over and caught fire. In his endeavours to extinguish the flames he received the injuries which terminated fatally. Applied science no less than pure science has had its martyrs, and among these Mansfield must be ranked.

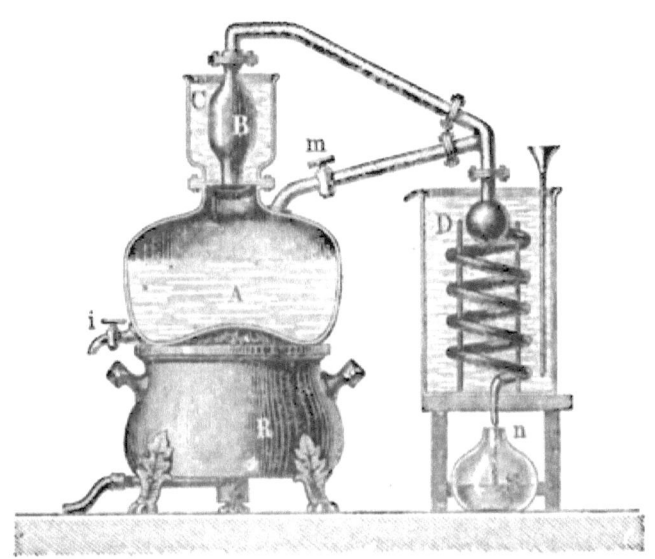

FIG. 5.—Mansfield's still. R the heating burner, A the body of the still with stopcock, i, for running out the contents. B the still-head kept in a cistern, C, of hot water or other liquid. The vapour generated by the boiling of the liquid in A, partly condenses in B, from whence the higher boiling-point portion flows back into the still. The uncondensed vapour passes into the condensing-worm, D, which is kept cool by a stream of water, and from thence flows into the receiver S. By opening m in the side-pipe any higher boiling-point oil condensing in the delivery-pipe can be run back into the still.

The operation of tar-distilling is about as unromantic a process as can be imagined, but it must be briefly described before the subsequent developments of the industry can be appreciated properly.

It has already been explained that the tar is a complex mixture of many different substances. These various compounds boil at certain definite temperatures, the boiling-point of a chemical compound being an inherent property. If a mixture of substances boiling at different temperatures is heated in a suitable vessel the compounds distil over, broadly speaking, in the order of their boiling-points. The separation by this process is not absolute, because compounds boiling at a certain temperature have a tendency to bring over with them the vapours of other compounds which boil at a higher temperature. But for practical purposes it will be sufficient to consider that the general tendency is for the compounds of low boiling-point to come over first, then the compounds of higher boiling-point, and finally those of the highest boiling-point. This is the principle made use of by the tar-distiller. The tar-still is a large iron pot provided with a still-head from which the vapours boil out into a coil of iron pipe kept cool in a vessel of water (see Fig. 6). The still is heated by a fire beneath it, and the different portions which condense in the iron coil are received in vessels which are changed as the different fractions of the tar come over. The process is what chemists would call a rough fractional distillation. The first fractions are liquid at ordinary temperatures, and the

AND WHAT WE GET FROM IT. 79

water in the condenser is kept cold; then, as the boiling-point rises, the fraction contains a hydrocarbon which solidifies on cooling, and the water in

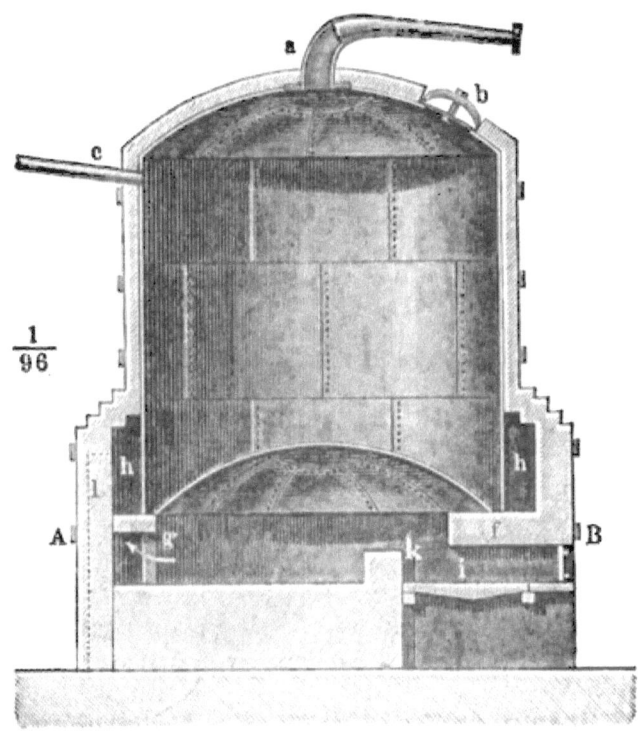

Fig. 6.—Sectional diagram of tar-still with arched bottom. The fireplace is at i; the hot gases pass over the bridge k and through g into the flues h, h. The pipe at c is to supply the still with tar; a is the exit pipe connected with the condenser, and b a man-hole for cleaning out the still. The condenser and bottom pipe for drawing off pitch have been omitted to avoid complication.

the condenser is made hot to prevent the choking up of the coil. Every one of these fractions of coal-tar, from the beginning to the end of the process, has its story to tell—all the chief constituents of the tar separated by this means have by chemical science been converted into useful products.

It is customary at the present time to collect four distinct fractions from the period when the tar begins to boil quietly, *i. e.* from the point when the small quantity of watery liquor which is unavoidably entangled with the tar has distilled over, by which time the temperature in the still is about 110° C. The small fraction that comes over up to this temperature constitutes what the tar-distiller calls "first runnings." From 110° C. to 210° C. there comes over a limpid inflammable liquid known as "light oil," and this is succeeded by a fraction which shows a tendency to solidify on cooling, owing to the separation of a solid crystalline hydrocarbon known as naphthalene. This last fraction, boiling between 210° C. and 240° C., is known as "carbolic oil," because it contains, in addition to the naphthalene, the chief portion of the carbolic acid present in the tar. From 240° C. to 270° C. there comes over another fraction which shows but little tendency to solidify in the condensing coil, and which is known as "heavy oil," or "creosote oil." From 270° C. up to the end of the distillation there distils a fraction which is viscid in consistency, and has a tendency to solidify on cooling owing to the separation of another crystalline hydrocarbon known as anthracene, and which gives the name of "anthracene oil" to this last fraction. When the latter has been collected there remains in the still the black viscid substance known as pitch, which is obtained of any desired consistency by leaving more or less of the anthracene oil mixed with it, or by afterwards mixing it with the heavy oil from previous distillations. The process carried out in the tar-still thus separates the tar into—(1) First runnings, up to 110° C. (2) Light oil, from 110° to 210° C. (3) Carbolic oil, from 210° to 240° C. (4) Creosote oil, from 240° to 270° C. (5) Anthracene oil, from 270° to pitch. (6) Pitch, left in still.

It has already been said that coal-tar is a complex mixture of various distinct chemical compounds. Included among the gases, ammoniacal liquor, and tar, the compounds which are known to be formed by the destructive distillation of coal already reach to nearly one hundred and fifty in number. Of the substances present in the tar, about a dozen are utilized as raw materials by the manufacturer, and these are contained in the fractions described above. The first runnings and light oil contain a series of important hydrocarbons, of which the three first members are known to chemists as benzene, toluene, and xylene, the latter being present in three different modifications.

F

The carbolic oil furnishes carbolic acid and naphthalene, and the anthracene oil the hydrocarbon which gives its name to that fraction. Here we have only half a dozen distinct chemical compounds to deal with, and if we confine our attention to these for the present, we shall be enabled to gain a good general idea of what chemistry has done with these raw materials. The products separated during the processes which have to be resorted to for the isolation of these raw materials have also their uses, which will be pointed out incidentally.

Beginning with the first runnings and the light oil, from which the hydrocarbons of the benzene series are separated, we have to make ourselves acquainted with the treatment to which these fractions are submitted by the tar-distiller. The light oil is first distilled from an iron still, similar to a tar-still, and the first portions which come over are added to the oily fraction brought over by the water of the first runnings. The separation of the oil from the water in this last fraction is a simple matter, because the hydrocarbons float as a distinct layer on the water, and do not mix with it. We have at this stage, therefore, four products to consider, viz. 1st, the oil from the first runnings; 2nd, the first portions of the light oil; 3rd, the later portions of the light oil; and 4th, the residue in the still. The first and second are mixed together, and the third is washed alternately with alkali and acid to remove acid and basic impurities, and can then be mixed with the first and second products. The total product is then ready for the next operation. The last portion of the light oil which remains in the still is useless as a source of benzene hydrocarbons, and goes into the heavy oil of the later tar fractions.

The process of purification is thus far one of fractional distillation combined with chemical washing. In fact, all the processes of purification to which these oils are submitted are essentially of the same character. The principle of fractional distillation has already been explained sufficiently for our present purpose. The process of washing a liquid may appear mysterious to the uninitiated, but in principle it is extremely simple. If we pour some water into a bottle, and then add some liquid which does not mix with the water—say paraffin oil—the two liquids form distinct layers, the one floating on the other. On shaking the bottle so as to mix the contents, the two liquids form a homogeneous mixture at first, but on standing for a short time separation into two layers again takes place. Now if there was present in the paraffin oil some substance soluble in the oil, but more soluble in water, such as alcohol, we should by the operation described wash the alcohol out of the oil, and when the liquids separated into layers after agitation the watery layer would contain the alcohol. By drawing off the oil or the water the former would then be obtained free from alcohol—it would have been "washed." This operation is precisely what the manufacturer does on a large scale with the coal-tar oils. These oils contain certain impurities of which some are of an acid character and dissolve in alkalies, while others are basic and dissolve in acid. The oil is therefore agitated in a suitable vessel provided with mechanical stirring gear with an aqueous solution of caustic soda, and after separation into layers the alkaline solution retaining the acid impurities is drawn off. Then the oil may be washed with water in the same way to remove the lingering traces of alkali, and then with acid—sulphuric acid or oil of vitriol—which dissolves out basic impurities and certain hydrocarbons not belonging to the benzene series which it is desirable to get rid of. A final washing with water removes any acid that may be retained by the oil.

The total product containing the benzene hydrocarbons is put through such a series of washing operations as above described, and is then ready for separation into its constituents by another and more perfect process of fractional distillation. This final separation is effected in a piece of apparatus somewhat complicated in structure, but simple in principle. It is a development on a large scale of the apparatus used by Mansfield in his early experiments. The details of construction are not essential to the present treatment of the subject, but it will suffice to say that the vapours of the boiling hydrocarbons ascend through upright columns, in which the compounds of high boiling-point first condense and run back into the still, while the lower boiling-point compounds do not condense in the columns, but pass on into a separate condenser, where they liquefy and are collected. But even with this rectification we do not get a perfect separation—the hydrocarbons are not perfectly pure from a chemical point of view, although they are pure enough for manufacturing purposes. Thus the first fraction consists of benzene containing a small percentage of toluene, then comes over a mixture containing a larger proportion of toluene, then comes a purer toluene mixed with a small percentage of xylene. The boiling-points of the three hydrocarbons are $81°$ C., $111°$ C., and $140°$ C. respectively; but owing to the nature of fractional distillation, a compound of a certain boiling-point always brings over with it a certain quantity of the compound of higher boiling-point, and that is why the rectifying column effects only a partial separation.

Of the hydrocarbons thus separated, benzene and toluene are by far the most important; there is but a limited use for the xylenes at present, and these and the hydrocarbons of higher boiling-point belonging to the same series which distil over between 140° and 150° constitute what is known as "solvent naphtha," because it is used for dissolving india-rubber for waterproofing purposes. The hydrocarbons of still higher boiling-point which remain in the still are used as burning naphtha for lamps. If benzene of a higher degree of purity is required—as it is for the manufacture of certain colouring-matters—the fraction containing this hydrocarbon can be again distilled through the rectifying column, and a large proportion of the toluene thus separated from it. Finally, pure benzene can be obtained by submitting the rectified hydrocarbon to a process of refrigeration in a mixture of ice and salt, when the benzene solidifies to a white crystalline solid, while the toluene does not solidify, and can be drained away from the benzene crystals which liquefy at about 5° C.

The account rendered by the technologist with respect to the light oils of the tar is thus a pretty good one. Already we see that benzene, toluene, solvent naphtha, and burning naphtha are separated from them. Even the alkaline and acid washings may be made to surrender their contained products, for the first of these contains a certain quantity of carbolic acid, and the acid contains a strongly smelling base called pyridine, for which there is at present no great demand, but which may one day become of importance. The actual quantity of benzene in tar is a little over one per cent. by weight, and of toluene there is somewhat less. The naphthas are present to the extent of about 35 per cent.

Now let us consider some of the transformations which benzene and toluene undergo in the hands of the manufacturing chemist. The production of aniline from benzene by acting upon this hydrocarbon with nitric acid, and then reducing the nitrobenzene, has already been referred to. For this purpose we now heat the nitrobenzene with iron dust and a little hydrochloric (muriatic) acid, and then distil over the aniline by means of a current of steam blown through the still. By a similar process toluene is converted into nitrotoluene, and the latter into toluidine.

The large quantity of aniline and toluidine now made has opened up a channel for the use of the waste borings from cast-iron. These are ground to a fine powder under heavy mill-stones, and constitute a most valuable reducing agent, known technically as "iron swarf." The metallic iron introduced in this form into the aniline still is converted into an oxide of iron by the action of the nitrobenzene, and this oxide of iron is used by the gas-maker for purifying the gas from sulphur as already described. When the oxide of iron is exhausted, *i.e.* when it has taken up as much sulphur as it can, it goes to the vitriol-maker to be burnt as a source of this acid. Here we have a waste product of the aniline manufacture utilized for the purification of coal-gas, and finally being made to give up the sulphur, which it obtained primarily from the coal, for the production of sulphuric acid, which is consumed in nearly every branch of chemical industry.

Nitrotoluene and toluidine each exist in three distinct modifications, so that it is more correct to speak of the nitrotoluenes and the toluidines; but the explanation of these differences belongs to pure chemical theory, and cannot now be attempted in detail. It must suffice to say that many compounds having the same chemical composition differ in their properties, and are said to be "isomeric," the isomerism being regarded as the result of the different order of arrangement of the atoms within the molecule.

Consider a homely illustration. A child's box of bricks contains a certain number of wooden blocks, by means of which different structures can be built up. Supposing all the bricks to be employed for every structure erected, the latter must in every case contain all the blocks, and yet the result is different, because in each structure the blocks are arranged in a different way. The bricks represent atoms, and the whole structure represents a molecule; the structures all have the same ultimate composition, and are therefore isomeric. This will serve as a rough analogy, only it must not be understood that the different atoms of the elements composing a molecule are of different sizes and shapes; on this point we are as yet profoundly ignorant.

Now as long ago as 1856, at the time when Perkin began making mauve by oxidizing aniline with bichromate of potash, it was observed by Natanson, that when aniline was heated with a certain oxidizing agent a red colouring-matter was produced. The same fact was observed in 1858 by Hofmann, who used the tetrachloride of carbon as an oxidizing agent. These chemists obtained the red colouring-matter as a by-product; it was formed only in small quantity, and was regarded as an impurity. In the same year, 1858, two French manufacturers patented the production of a red dye formed by the action of chromic acid and other oxidizing agents on aniline, the colouring matter thus made being used for dying artificial flowers. Then, a year later, the French chemist

Verguin found that the best oxidizing agent was the tetrachloride of tin, and this with many other oxidizing substances was patented by Renard Frères and Franc, and under their patent the manufacture of the aniline red was commenced on a small scale in France. Finally, in 1860, an oxidizing agent was made use of almost simultaneously by two English chemists, Medlock and Nicholson, which gave a far better yield of the red than any of the other materials previously in use, and put the manufacture of the colouring-matter on quite a new basis. The oxidizing material patented by Medlock and Nicholson is arsenic acid, and their process is carried on at the present time on an enormous scale in all the chief colour factories in Europe, the colouring-matter produced by this means being generally known as fuchsine or magenta.

In four years the accidental observation of Natanson and Hofmann, made, be it remembered, in the course of abstract scientific investigation, had thus developed into an important branch of manufacture. A demand for aniline on an increased scale sprung up, and the light oils of coal-tar became of still greater importance. The operations of the tar-distiller had to undergo a corresponding increase in magnitude and refinement; the production of nitrobenzene and necessarily of nitric acid had to be increased, and a new branch of manufacture, that of arsenic acid from arsenious acid and nitric acid, was called into existence. Perkin's mauve prepared the way for the manufacture of aniline, and the discovery of a good process for the production of magenta increased this branch of manufacture to a remarkable extent. Still later in the history of the magenta manufacture, attempts were made, with more or less success, to use nitrobenzene itself as an oxidizing agent, and a process was perfected in 1869 by Coupier, which is now in use in many factories.

The introduction of magenta into commerce marks an epoch in the history of the coal-tar colour industry—pure chemistry and chemical technology both profited by the discovery. The brilliant red of this colouring-matter is objected to by modern æstheticism, but the dye is still made in large quantities, its value having been greatly increased by a discovery made about the same time by John Holliday and the Baden Aniline and Soda Company, and patented by the latter in 1877. Magenta is the salt of a base now known as rosaniline, and it belongs therefore to the class of basic colouring-matters. The dyes of this kind are as a group less fast, and have a more limited application than those colouring-matters which possess an acid character, so that the discovery above referred to—that magenta could be converted into an acid without destroying its colouring power by acting upon it with very strong sulphuric acid—opened up a new field for the employment of the dye, and greatly extended its usefulness. In this form the colouring-matter is met with under the name of " acid magenta."

It must be understood that the production of magenta from aniline by the oxidizing action of arsenic acid or nitrobenzene is the result of chemical change; the colouring-matter is no more present in the aniline than the latter is contained in the benzene. And just in the same way that the colourless aniline oil by chemical transformation gives rise to the intensely colorific magenta, so the latter by further chemical change can be made to give rise to whole series of different colouring-matters, each consisting of definite chemical compounds as distinct in individuality as magenta itself. Thus in 1860, about the time when the arsenic acid process was inaugurated, two French chemists, Messrs. Girard and De Laire, observed that by heating rosaniline for some time with aniline and an aniline salt, blue and violet colouring-matters were produced. This observation formed the starting-point of a new manufacture proceeding from magenta as a raw material. The production of the new colouring-matters was perfected by various investigators, and a magnificent blue was the final result. But here also the dye was of a basic character, and being insoluble in water had only a limited application, as a spirit bath had to be used for dissolving the substance. In 1862, however, an English technologist, the late E. C. Nicholson, found that by the action of strong sulphuric acid the aniline blue could be rendered soluble in water or alkali, and the value of the colouring-matter was enormously increased by this discovery. The basic and slightly soluble spirit blue was by this means converted into acid blues, which are now made in large quantities, and sold under the names of Nicholson's blue, alkali blue, soluble blue, and other trade designations. There is at the present time hardly any other blue which for fastness, facility of dyeing, and beauty can compete with this colouring-matter introduced by Nicholson as the outcome of the work of Girard and De Laire.

Other transformations of rosaniline have yet to be chronicled. In 1862 Hofmann found that by acting upon this base—the base of magenta—with the iodide of methyl, violet colouring-matters were produced, and these were for some years extensively employed under the name of Hofmann's violets. And still more remarkable, by the prolonged action of an excess of methyl iodide upon rosaniline, Keisser found that a green colouring-matter was formed. The latter was patented in 1866, and the dye was for some time in use under the name of "iodine green." The statement that technology profited by the introduction of magenta has therefore been justified.

It remains to add, that the tar obtained from one ton of Lancashire coal furnishes an amount of aniline capable of giving a little over half a pound of magenta. The colouring power of the latter will be inferred from the fact, that this quantity would dye 375 square yards of white flannel of a full red colour, and if converted into Hofmann violet by methylation, would give enough colour to dye double this surface of flannel of a deep violet shade. It should be stated also, that during the formation of magenta by the arsenic acid process, there are formed small quantities of other colouring-matters which are utilized by the manufacturer. Among these by-products is a basic orange dye, which was isolated by Nicholson, and investigated by Hofmann in 1862. Under the name of "phosphine" this colouring-matter is still used, especially for the dyeing of leather. Even the spent arsenic acid of the magenta-still has its use. The arsenious acid resulting from the reduction of this arsenic acid is generally obtained in the form of a lime salt after the removal of the magenta by the purifying processes to which the crude product is submitted. From the arsenical waste arsenious acid can be recovered, and converted back into arsenic acid by the action of nitric acid. Quite recently the arsenical residue has been used with considerable success in America as an insecticide for the destruction of pests injurious to agricultural crops.

Concurrently with these technical developments of coal-tar products, the scientific chemist was carrying on his investigations. The compounds which science had given to commerce were made on a scale that enabled the investigator to obtain his materials in quantities that appeared fabulous in the early days when aniline was regarded as a laboratory curiosity, and magenta had been seen by only a few chemists.

The fundamental problem which the modern chemist seeks to solve is in the first place the composition of a compound, *i.e.* the number of the atoms of the different elements which form the molecule, and in the next place the way in which these atoms are combined in the molecule. Reverting to our former analogy, the first thing to be found is how many different blocks enter into the composition of the structure, and the next thing is to ascertain how the blocks are arranged. When this is done, we are said to know the "constitution" or "structure" of the molecule, and in many cases when this is known we can build up or synthesise the compound by combining its different groups of atoms by suitable methods. The coal-tar industry abounds with such triumphs of chemical synthesis; a few of these achievements will be brought to light in the course of the remaining portions of this work.

The chemical investigation of magenta was commenced by Hofmann, whose name is inseparably connected with the scientific development of the coal-tar colour industry. In 1862 he showed that magenta was the salt of a base which he isolated, analysed, and named rosaniline. He established the composition of this base and of the violet and blue colouring-matters obtained from it by the processes already described. In 1864 he made the interesting discovery that magenta is not formed by the oxidation of *pure* aniline, but that a mixture of aniline and toluidine is essential for the production of this colouring-matter. In fact, the aniline oil used by the manufacturer had from the beginning consisted of a mixture of aniline and toluidine, and at the present time "aniline for red" is made by nitrating a mixture of benzene and toluene and reducing the nitro-compounds.

From this work of Hofmann's suggestions naturally arose concerning the "constitution" of rosaniline, and new and fruitful lines of work were opened up. Large numbers of chemists of the greatest eminence pursued the inquiry, but the details of their work, although of absorbing interest to the chemist, cannot be discussed in the present volume. The final touch to a long series of investigations was given by two German chemists, Emil and Otto Fischer, who in 1878 proved the constitution of rosaniline by obtaining from it a hydrocarbon, the parent hydrocarbon from which the colouring-matter is derived. The purely scientific discovery of the Fischers threw a flood of light on the chemistry of magenta, and enabled a large number of colouring-matters related to the latter to be classed under one group, having the parent hydrocarbon as a central type. This hydrocarbon, it may be remarked, is known as triphenylmethane, as it is a derivative of methane, or marsh gas. The blues and violets obtained from rosaniline belong to this group, and so also do certain other colouring-matters which had been manufactured before the Fischers' discovery. In order to carry on the story of the utilisation of aniline, it is necessary to know something about these other colouring-matters which are obtained from it.

It has been explained that by the methylation of rosaniline Hofmann obtained violet colouring-matters. Now as rosaniline is obtained by the oxidation of a mixture of aniline and toluidine, it seems but natural that if these bases were methylated first and then oxidized a violet dye would be produced. The French chemist Lauth first obtained a violet colouring-matter by this method in 1861. In 1866 this violet dye was manufactured in France by Poirrier, and it is still made in large quantities, being known under the name of "methyl violet." This colouring-matter, and a bluer derivative of it discovered in 1868, gradually displaced the Hofmann violets, chiefly owing to their greater cheapness of production. We are thus introduced to methylated aniline as a source of colouring-matters, and as the compound in question has many different uses in the coal-tar industry, a few words must be devoted to its technology.

Aniline, toluidine, and similar bases can be methylated by the action of methyl iodide, but the cost of iodine is too great to enable this process to be used by the manufacturer. Methyl chloride, however, answers equally well, and this compound, which is a liquid of very low boiling point ($-23°$ C.), is prepared on a large scale from the waste material of another industry, viz. the beet-sugar manufacture. It is interesting to see how distinct industries by chemical skill are made to act and react upon one another. Thus the cultivation of the beet, as already explained, is largely dependent on the supply of ammonia from gas-liquor. During the refining of the beet-sugar, a large quantity of uncrystallisable treacle is separated, and this is fermented for the manufacture of alcohol. When the latter is distilled off there remains a spent liquor containing among other things potassium salts and nitrogenous compounds. This waste liquor, called "vinasse," is evaporated down and ignited in order to recover the potash, and during the ignition, ammonia, tar, gas, and other volatile products are given off. Among the volatile products is a base called trimethylamine, which is a derivative of ammonia; the salt formed by combining trimethylamine with hydrochloric acid when heated gives off methyl chloride as a gas which can be condensed by pressure.

Here we have a very pretty cycle of chemical transmigration. The nitrogen of the coal plants, stored up in the earth for ages, is restored in the form of ammonia to the crops of growing beet; the nitrogen is made to enter into the composition of the latter plant by the chemico-physiological process going on, and the nitrogenous compounds removed from the plant and heated to the point of decomposition in presence of the potash (which also entered into the composition of the plant), give back their nitrogen partly in the form of a base from which methyl chloride can be obtained. The latter is then made to methylate a product, aniline, derived indirectly from coal-tar. The utilisation of the "vinasse" for this purpose was made known by Camille Vincent of Paris in 1878.

The methylation of aniline can obviously be carried out by the foregoing process only when beet-sugar residues are available. There is another method which is more generally used, and which is interesting as bringing in a distinct branch of industry. The same result can, in fact, be arrived at by heating dry aniline hydrochloride, *i.e.* the hydrochloric acid salt of aniline, with methyl alcohol or wood-spirit in strong metallic boilers under great pressure. This is the process carried on in most factories, and it involves the use of pure methyl alcohol, a branch of manufacture which has been called into existence to meet the requirements of the coal-tar colour maker.[1] This alcohol or wood-spirit is obtained by the destructive distillation of wood, and is purified by a series of operations which do not at present concern us. It must be mentioned that the product of the methylation of aniline, which it is the object of the manufacturer to obtain, is an oily liquid called dimethylaniline, which, by

[1] This remark does not apply to Great Britain; our Excise regulations have practically killed those branches of manufacture requiring the use of pure wood-spirit.

virtue of the chemical transformation, is quite different in its properties to the aniline from which it is derived. By a similar operation, using ethyl alcohol, or spirit of wine, diethylaniline can be obtained, and by heating dry aniline hydrochloride with aniline under similar conditions a crystalline base called diphenylamine is also prepared.

Now these products—dimethylaniline, diethylaniline, and diphenylamine—are derived from aniline, and they are all sources of colouring-matters. Methyl-violet is obtained by the oxidation of dimethylaniline by means of a gentle oxidizer; a mixture of bases is not necessary as in the case of the magenta formation. Then in 1866 diphenylamine was shown by Girard and De Laire to be capable of yielding a fine blue by heating it with oxalic acid, and this blue, on account of the purity of its shade, is still an article of commerce. It can be made soluble by the action of sulphuric acid in just the same way as the other aniline blue. Furthermore, by acting with excess of methyl chloride on methyl violet, a brilliant green colouring-matter was manufactured in 1878, which was obviously analogous to the iodine green already mentioned, and which for some years held its own as the only good coal-tar green. These are the dyes—methyl violet and green, and diphenylamine blue—which were in commerce before the discovery of the Fischers, and which this discovery enabled chemists to class with magenta, aniline blue, and Hofmann violet in the triphenylmethane group.

Later developments bring us into contact with other dyes of the same class, and with the industrial evolution of the purely scientific idea concerning the constitution of the colouring-matters of this group. Benzene and toluene again form the points of departure. By the action of chlorine upon the vapour of boiling toluene there are obtained, according to the extent of the action of the chlorine, three liquids of use to the colour manufacturer. The first of these is benzyl chloride, the second benzal chloride, and the third benzotrichloride or phenyl chloroform. Benzyl chloride, it may be remarked in passing, plays the same part in organic chemistry as methyl chloride, and enables certain compounds to be benzylated, just in the same way that they can be methylated. The bluer shade of methyl violet, introduced in 1868, and still manufactured, is a benzylated derivative. By the action of benzotrichloride on dimethylaniline in the presence of dry zinc chloride, Oscar Doebner obtained in 1878 a brilliant green colouring-matter which was manufactured under the name of "malachite green." It will be remembered that this was about the time when the Fischers were engaged with their investigations. These last chemists, by virtue of their scientific results, were enabled to show that Doebner's green was a member of the triphenyl-methane group, and they prepared the same compound by another method which has enabled the manufacturer to dispense with the use of the somewhat expensive and disagreeable benzotrichloride. The Fischers' method consists in heating dimethyl-aniline with bitter-almond oil and oxidizing the product thus formed, when the green colouring-matter is at once produced. This method brings the technologist into competition with Nature, and we shall see the result.

Benzoic aldehyde or bitter-almond oil is one of the oldest known products of the vegetable kingdom, and has from time to time been made the subject of investigation by chemists since the beginning of the century. It arises from the fermentation of a nitrogenous compound found in the almond, and known as amygdalin, the nature of the fermentative change undergone by this substance having been brought to light by Wöhler and Liebig. The discovery of a green dye, requiring for its preparation a vegetable product which was very costly, compelled the manufacturer to seek another source of the oil. Pure chemistry again steps in, and solves the problem. In 1863 it was known to Cahours that benzal chloride, on being heated with water or alkali, gave benzoic aldehyde, and in 1867 Lauth and Grimaux showed that the same compound could be formed by oxidizing benzyl chloride in the presence of water. It was but a step from the laboratory into the factory in this case, and at the present time the aldehyde is made on a large scale by chlorinating boiling toluene beyond the stage of benzyl chloride, and heating the mixture of benzal chloride and benzotrichloride with lime and water under pressure. By this means the first compound is transformed into benzoic aldehyde, and the second into benzoic acid. This last substance is also required by the colour-maker, as it is used in the manufacture of blue by the action of aniline on rosaniline; without some such organic acid the transformation of rosaniline into the blue is very imperfect.

Benzoic acid, like the aldehyde, is a natural product which has long been known. It was obtained from gum benzoïn at the beginning of the seventeenth century, and its preparation from this source was described by Scheele in 1755. The same chemist afterwards found it in urine, and from these two sources, the one vegetable and the other animal, the acid was formerly prepared. Its relationship to benzene has already been alluded to in connection with the history of that hydrocarbon. It will be remembered that by heating this acid with lime Mitscherlich obtained benzene in 1834. In one operation, therefore, setting out from toluene, we make these two natural products, the aldehyde and acid, which are easily separable by technical processes. The wants of the technologist have been met, and he has been enabled to compete successfully with Nature, for he can manufacture these products much more cheaply than when he had to depend upon bitter almonds or gum benzoïn. The synthetical bitter-almond oil is chemically identical with that from the plant. Besides its use for the manufacture of colouring-matters, it is employed for flavouring purposes and in perfumery, this being the first instance of a coal-tar perfume which we have had occasion to mention. The odour in this case, it must be remembered, is that of the actual compound which imparts the characteristic taste and smell to the almond ; it is not the result of substituting a substance which has a particular odour for another having a similar odour, as is the case with nitrobenzene, which, as already mentioned, is used in large quantities under the name of "essence of mirbane," for imparting an almond-like smell to soap.

The introduction of malachite green marks another epoch in the history of the technology of the triphenylmethane colours. The action between benzoic aldehyde and other bases analogous to dimethylaniline was found to be quite general, and the principle was extended to diethylaniline and similarly constituted bases. Various green dyes—some of them acids formed by the action of sulphuric acid on the colour base—are now manufactured, and many other colouring-matters of the same group are synthesised by the benzoic aldehyde process.

One other development of this branch of manufacture has yet to be recorded. The new departure was made in 1883 by Caro and Kern, who patented a process for the synthesis of colouring-matters of this group. In this synthesis a gas called phosgene is used, the said gas having been discovered by John Davy in 1811, who gave it its name because it is formed by the direct union of chlorine and carbon monoxide under the influence of sunlight. Caro and Kern's process is the first technical application of Davy's compound. By the action of phosgene on dimethylaniline and analogous bases in the presence of certain compounds which promote the chemical interaction, a number of basic colouring-matters of brilliant shades of violet ("crystal violet") and blue ("Victoria blue," "night blue") are produced, these being all members of the triphenylmethane group. One of these dyes is a fine basic yellow known as "auramine," which is a derivative of diphenylmethane.

AND WHAT WE GET FROM IT. 107

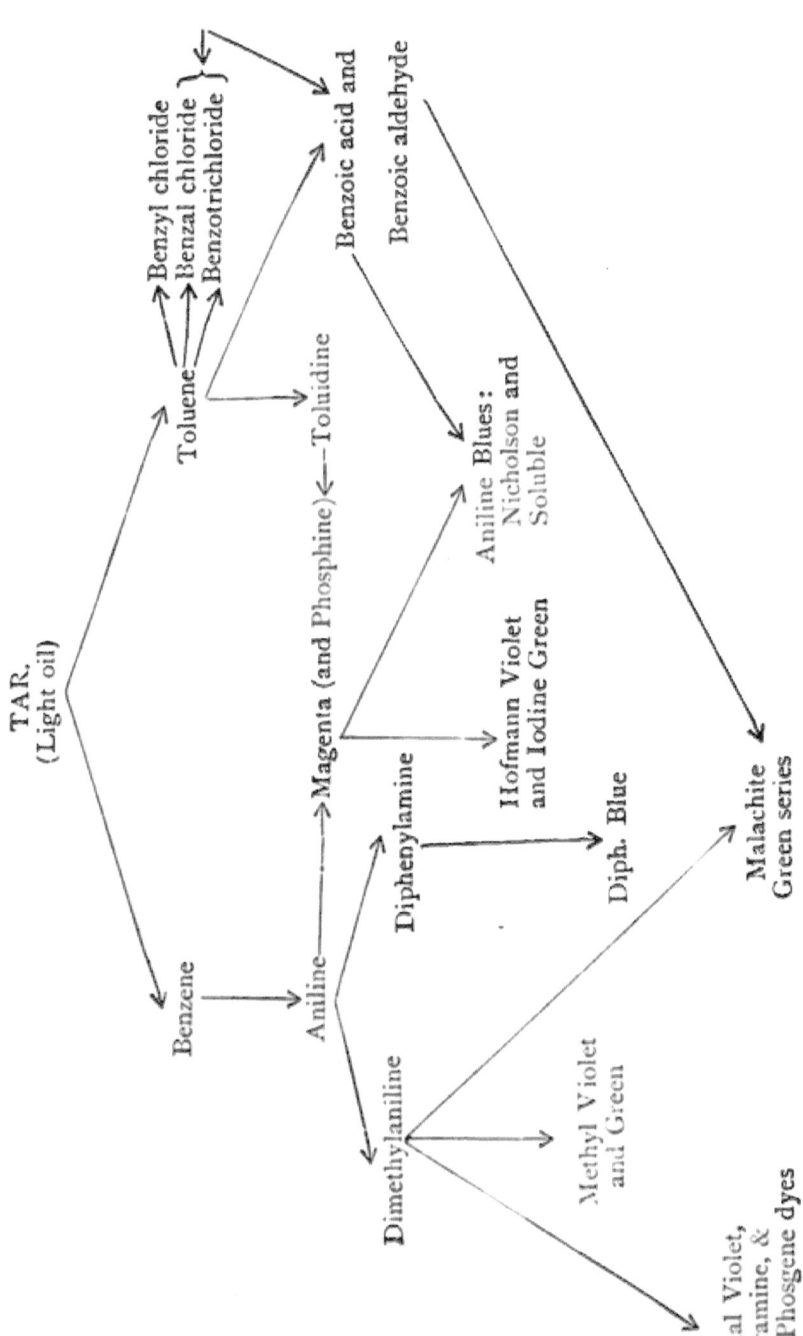

From benzene and toluene alone about forty distinct colouring-matters of the rosaniline group are sent into commerce. The relationship of those compounds to each other and to their generating substances is not easy to grasp by those to whom the facts are presented for the first time. The scheme on page 107 shows these relationships at a glance.

The colouring-matters derived from these two hydrocarbons are far from being exhausted. During the oxidation of aniline for the production of mauve—which colouring-matter, it may be mentioned, is no longer made—a red compound is formed as a by-product. This was isolated by Perkin in 1861, and studied scientifically by Hofmann and Geyger, who established its composition in 1872, the dye being at that time manufactured under the name of "saffranine." It appears to have been first introduced about 1868. The conditions of formation of this dye were at first imperfectly understood, but the problem was attacked by chemists and technologists, and the first point of importance resulting from their work was that saffranine was derived from one of the toluidines present in the commercial aniline. To record the various steps in this chapter of industrial chemistry would take us beyond the scope of the present work. In addition to the chemists named,

Caro, Bindschedler, and others contributed to the technology, while the scientific side of the matter was first taken up by Nietzki in 1877, by Otto Witt in 1878, and by Bernthsen in 1886. It is to the work of these chemists, and especially to that of Witt, that we owe our present knowledge of the constitution of this and allied colouring-matters. Space will not admit of our traversing the ground, although to chemists it is a line of investigation full of interest; it will be sufficient to say that by 1886 these investigators had accomplished for these colouring-matters what the Fischers had done for the rosaniline group—they established their constitution, and showed that they were derivatives of diphenylamine, containing two nitrogen atoms joined together in a particular way. The parent-substance from which these compounds are derived is known at the present time as "azine" (French, azote = nitrogen), and the dyes belong accordingly to the azine group. The first coal-tar colouring-matter, Perkin's mauve, is a member of this class.

The azine dyes are basic, and mostly of a red or pink shade; they are somewhat fugitive when exposed to light, but possess a certain value on account of their affinity for cotton, and the readiness with which they can be used in admixture with other colouring-matters. Some of the best known are made by oxidizing certain derivatives of aniline or toluidine, in the presence of these or analogous bases. To make this intelligible a little more chemistry is necessary. Aniline is a derivative of benzene in which one atom of hydrogen is replaced by the residue of ammonia. Ammonia is composed of one atom of nitrogen and three atoms of hydrogen; benzene is composed of six atoms of carbon and six of hydrogen. If one atom of hydrogen is supposed to be withdrawn from ammonia, there remains a residue called the amido-group, and if we imagine this group to be substituted for one of the hydrogen atoms in benzene, we have an amido-derivative, *i. e.* amido-benzene or aniline. Similarly, the toluidines are amidotoluenes. If two hydrogen atoms in benzene or toluene are replaced by two amido-groups, we have diamidobenzenes and diamidotoluenes, which are strongly basic substances, capable of existing in several isomeric modifications. Certain of these diamido-compounds when oxidized in the presence of a further quantity of aniline, toluidine, and such amido-compounds, give rise to unstable blue products, which readily become transformed into red dyes of the azine group.

Some azine dyes are produced by another method, which is instructive because it brings us into contact with a derivative of dimethylaniline which figures largely in the coal-tar colour industry. By the action of nitrous acid on this base, there is produced a compound known as nitrosodimethylaniline, which was discovered by Baeyer and Caro in 1874, and which contains the residue of nitrous acid in place of one atom of hydrogen. The residue of nitric acid which replaces hydrogen in benzene is the nitro-group, and the compound is nitrobenzene. The analogy with nitrous acid will therefore be sufficiently understood—the residue of this acid is the nitroso-group, and compounds containing this group are nitroso-derivatives. In 1879, Otto Witt found that the nitroso-group in nitrosodimethylaniline acted as an oxidizing group, and enabled this compound to act upon certain diamido-derivatives of benzene and toluene, with the formation of unstable blue compounds, which on heating the solution changed into red colouring-matters of the azine group. This process soon bore fruit industrially, and azines of a red, violet, and blue shade were introduced under the names of neutral red, violet, and blue, Basle blue, &c., some of these surviving at the present time.

We have now to turn to another chapter in the history of dimethylaniline. In 1876, Lauth discovered a new colour test for one of the diamido-benzenes. By heating this base with sulphur, and oxidizing the product, a violet colouring-matter was formed, and the same compound was produced by oxidizing the base in an aqueous solution in the presence of sulphuretted hydrogen. Lauth's violet was never manufactured in quantity because the yield is small; but in the hands of Dr. Caro the work of Lauth bore fruit in another direction. Instead of using the diamidobenzene, Caro used its dimethyl-derivative, and by this means obtained a splendid blue dye, which was introduced under the name of "methylene blue." Here again we find scientific research reacting on technology. A few words of chemical explanation will make this manufacture intelligible. By the action of reducing agents on nitro and nitroso-compounds, the nitro and nitroso-group become converted into the amido-group. Thus when nitrobenzene is reduced by iron and an acid we get aniline; similarly when nitrosodimethylaniline is reduced by zinc and an acid we get amidodimethylaniline, and this is the base used in the preparation of methylene blue. By oxidizing this base in the presence of sulphuretted hydrogen, the colouring-matter is formed. Other methods of arriving at the same result were discovered and patented in due course, but the various processes cannot be discussed here.

Lauth's violet and methylene blue became the subjects of scientific investigation in 1879 by Koch, and in 1883 a series of brilliant researches were commenced by Bernthsen which extended over several years, and which established the constitution of these compounds. It was shown that they are derivatives of diphenylamine containing sulphur as an essential constituent. The parent-compound is diphenylamine in which sulphur replaces hydrogen, and is therefore known as thiodiphenylamine. It can be prepared by heating diphenylamine with sulphur, and is sometimes called thiazine, because it is somewhat analogous in type to azine. We must therefore credit dimethylaniline with being the industrial generator of the thiazines. The blue is largely used for cotton dyeing, producing on this fibre when properly mordanted an indigo shade. By the action of nitrous acid the blue is converted into a green known as "methylene green."

Although the scope of this work admits of our dealing with only a few of the more important groups of colouring-matters, it will already be evident that the chemist has turned benzene and toluene to good account. But great as is the demand for these hydrocarbons for the foregoing purposes, there are other branches of the coal-tar industry which are dependent upon them. It will serve as an answer to those who are continually raising the cry of brilliancy as an offence to æsthetic taste if we consider in the next place a most valuable and important black obtained from aniline. All chemists who studied the action of oxidizing agents, such as chromic acid, on aniline, from Runge in 1834 to Perkin in 1856, observed the formation of greenish or bluish-black compounds. After many attempts to utilize these as colouring-matters, success was achieved by John Lightfoot of Accrington near Manchester in 1863. By using as an oxidizing agent a mixture of potassium chlorate and a copper salt, Lightfoot devised a method for printing and dyeing cotton fabrics, the use of which spread rapidly and created an incieased demand for the hydrochloride of aniline, this salt being now manufactured in enormous quantities under the technical designation of "aniline salt." Lightfoot's process was improved for printing purposes by Lauth in 1864, and many different oxidizing mixtures have been subsequently introduced, notably the salts of vanadium, which are far more effective than the salts of copper, and which were first employed by Lightfoot in 1872. In 1875-76 Coquillion and Goppelsröder showed that aniline black is produced when an electric current is made to decompose a solution of an aniline salt, the oxidizing agent here being the nascent oxygen resulting from the electrolysis. In these days when the generation of electricity is so economically effected, this process may become more generally used, and the coal-tar industry may thus be brought into relationship with another branch of applied science. Aniline black is seldom used as a direct colouring-matter; it is generally produced in the fibre by printing on the mixture of aniline salt and oxidizing compounds thickened with starch, &c., and then allowing the oxidation to take place spontaneously in a moist and slightly heated atmosphere. By a similar process, using a dye-bath containing the aniline salt and oxidizing mixture, cotton fibre is easily dyed. The black cannot be used for silk or wool, as the oxidizing materials attack these fibres, but for cotton dyeing and calico printing this colouring-matter has come seriously into competition with the black dyes obtained from logwood and madder. The use of aniline for this purpose, first rendered practicable by Lightfoot, is among the most important of the many wonderful applications of coal-tar products in the tinctorial industry.

The year 1863 witnessed the introduction of the first of a new series of colouring-matters which have had an enormous influence both on the art of the dyer as well as in the utilization of tar-products which were formerly of but little value. We can consider the history of some of these colours now, because the earliest of them was produced from aniline. The formation of a yellow compound when nitrous acid acts upon aniline was observed by several chemists prior to the date mentioned. In 1863 the firm of Simpson, Maule and Nicholson manufactured a yellow dye by passing nitrous gas into a solution of aniline in alcohol, and this had a limited application under the name of "aniline yellow." Soon afterwards, viz. in 1866, the firm of Roberts, Dale & Co. of Manchester introduced a brown dye under the name of "Manchester brown"—this compound, which was discovered by Dr. Martius in 1865, having been produced by the action of nitrous acid on one of the diamidobenzenes. Ten years later Caro and Witt discovered an orange colouring-matter belonging to the same class, and the latter introduced the compound into commerce as "chrysoïdine." These three compounds are basic, and the first of them is no longer used as a direct dye because it is fugitive. Chrysoïdine is still used to a large extent, and the brown—now known as "Bismarck brown"—is one of the staple products of the colour manufacturer at the present time. From this fragment of technological history let us now turn to chemical science.

The chemist whose name will always be associated with the compounds of this group is the late Dr. Peter Griess of Burton-on-Trent. He commenced his study of the action of nitrous acid on organic bases in 1858, and from that time till the period of his death in 1888, he was constantly contributing to our knowledge of the resulting compounds. In 1866, he and Dr. Martius established the composition of aniline yellow, and the following year Caro and Griess did the same thing for the Manchester brown. In 1877 Hofmann and Witt established the constitution of chrysoïdine, the final outcome of all this work being to show that the three colouring-matters belonged to the same group. The further development of these discoveries has been one of the most prolific sources of new colouring-matters. A brief summary of our present position with respect to this group must now be attempted.

When nitrous acid acts upon an amido-derivative of a benzenoid hydrocarbon in the presence of a mineral acid, there is formed a compound in which the amido-group is replaced by a pair of nitrogen atoms joined together in a certain way, which is different to the mode of combination in the azines. This pair of nitrogen atoms is combined on the one hand with the hydrocarbon residue, and on the other with the residue of the mineral acid. The resulting compound is very unstable; its solution decomposes very readily, and generally has to be kept cool by ice. Freezing machines turning out large quantities of ice are kept constantly at work in factories where these products are made. The latter are known as "diazo-compounds"— Griess's compounds *par excellence*—and they are prepared on a large scale by dissolving a salt of the amido-base, generally the hydrochloride, in water with ice, and adding sodium nitrite. The result is a diazo-salt; aniline, for example, giving diazobenzene chloride, and toluidine diazotoluene chloride. Similarly all amido-derivatives of a benzenoid character can be "diazotised." The importance of this discovery will be seen more fully in the next chapter. At present we are more especially concerned with aniline.

The extreme instability of the diazo-salts enables them to combine with the greatest ease with amido-derivatives and with other compounds. The very property which in the early days rendered their investigation so difficult, and which taxed the ingenuity of chemists to the utmost, has now placed these compounds in the front rank as colour generators. When a diazo-salt acts on an amido-derivative there is formed a compound which is more or less unstable, but which readily undergoes transformation under suitable conditions into a stable substance in which two hydrocarbon residues are joined together by the pair of nitrogen atoms. These products are dye-stuffs, known as "azo-colours," and aniline yellow, Bismarck brown, and chrysoïdine are the oldest known technical compounds belonging to the group. The parent substance is "azobenzene," and these three colouring-matters are mono-, di- and triamido-azobenzene respectively.

A new phase in the technology of tar-products was entered upon when Witt caused a diazo-salt to act upon diamidobenzene. This was the first industrial application of Griess's discovery. Azobenzene, which was discovered by Mitscherlich in 1834, and azotoluene are now manufactured by reducing nitrobenzene and nitrotoluene with mild reducing agents. These parent compounds are not in themselves colouring-matters, but they are transformed into bases which give rise to a splendid series of azo-dyes, as will be described subsequently. Let it be recorded here that these two compounds are to be added to the list of valuable products obtained from benzene and toluene. And it must also be remembered that the introduction of these azo-colours has necessitated the manufacture on a large scale of sodium nitrite as a source of nitrous acid. Without entering into unnecessary detail it may be stated broadly that this salt is made by fusing Chili saltpetre, which is the nitrate of sodium, with metallic lead, litharge or oxide of lead being obtained as a secondary product. Then again, the manufacture of Bismarck brown requires dinitrobenzene, this being made by the nitration of benzene beyond the stage of nitrobenzene. The brown is made by reducing the dinitrobenzene to diamidobenzene, and then treating a solution of the latter with sodium nitrite and an acid. The azo-colour is formed at once, and no special refrigeration is required in this particular case.

It has already been stated that the old aniline yellow of 1863 is no longer used on account of its fugitive character. In 1878 Grässler found that by the action of very strong sulphuric acid this azo-compound could be converted into a sulpho-acid in just the same way that magenta can be converted into acid magenta. Under the name of "acid yellow" this sulpho-acid is now used, not only as a direct yellow colouring-matter, but as a starting-point in the manufacture of other azo-dyes. The use of acid yellow for this last purpose will be dealt with again in the next chapter.

There is one other use for aniline yellow which dates from the year of its discovery, when Dale and Caro found that by adding sodium nitrite to aniline hydrochloride and heating the mixture, a blue colouring-matter is produced. The latter was introduced in 1864 under the name of "induline." It was shown subsequently by the scientific researches of several chemists that the blue produced by Dale and Caro's method results from the action of the aniline salt on the aniline yellow, which is formed by the action of the nitrous acid on the aniline and aniline salt. This explanation was proved to be correct in 1872 by Hofmann and Geyger, who prepared the colouring-matter by heating aniline yellow and aniline salt with alcohol as a solvent. These chemists established the composition and gave it the name of "azodiphenyl blue." Later, viz. in 1883, the manufacture was improved by Otto Witt and E. Thomas, and the dye, under the old name of "induline," is now largely manufactured by first preparing aniline yellow and then heating this with aniline and aniline salt. The colouring-matter as formed by this method is basic and insoluble in water; it is made acid and soluble by treatment with sulphuric acid, which converts it into a sulpho-acid. Induline belongs to the sober-tinted colours, and produces a shade somewhat resembling indigo. Closely related thereto is a bluish-grey called "nigrosine," obtained by heating nitrobenzene with aniline, as well as a certain bluish by-product obtained during the formation of magenta, and known as "violaniline."

It will be convenient here to pause and reflect upon the great industrial importance of the two coal-tar hydrocarbons upon which we have thus far concentrated our attention. Their uses are by no means exhausted as yet, but they have already been made to account for such a number of valuable products that the reader may find it useful to have the results presented in a collected form. This is given below as a chronological summary—

1856. Mauve discovered by Perkin; leading to manufacture of aniline.
1860. Arsenic acid process for magenta discovered; leading to manufacture of arsenic acid.
1860. Aniline blue discovered; leading in 1862 to soluble and Nicholson blues.
1861. Methyl violet discovered; manufactured in 1866; leading to a new use of copper salts as oxidizing agents, and to the manufacture of dimethylaniline.
1862. Hofmann violets discovered; leading to manufacture of methyl iodide from iodine, phosphorus, and wood spirit.
1862. Phosphine (chrysaniline) discovered in crude magenta.
1863. Aniline black introduced; leading to a new use for potassium chlorate and copper salts, and to the manufacture of aniline salt.
1863. Aniline yellow introduced, the first azo-colour.
1864. Induline discovered; leading to new use for aniline yellow.
1866. Manchester brown introduced, the second azo-colour; leading to the manufacture of sodium nitrite, and of dinitrobenzene.
1866. Iodine green introduced; leading to further use for methyl iodide.
1866. Diphenylamine blue introduced; leading to manufacture of diphenylamine.

1868. Blue shade of methyl violet introduced; leading to manufacture of benzyl chloride.
1868. Saffranine introduced.
1869. Nitrobenzene process for magenta discovered.
1876. Chrysoïdine introduced, the third azo-colour.
1876. Methylene blue introduced; leading to manufacture of nitrosodimethylaniline.
1877. Acid magenta discovered.
1878. Methyl green introduced; leading to utilization of waste from beet-sugar manufacture.
1878. Malachite green discovered; leading to manufacture of benzoic aldehyde.
1878. Acid yellow discovered; leading to new use for aniline yellow.
1879. Neutral red and allied azines introduced; leading to a new use for nitrosodimethylaniline.
1883. Phosgene colours of rosaniline group introduced; leading to manufacture of phosgene.

CHAPTER III.

AMONG the most venerable of natural dye-stuffs is indigo, the substance from which Unverdorben first obtained aniline in 1826. The colouring matter is found in a number of leguminous (see Fig. 7), cruciferous, and other plants, some of which are largely cultivated in India, China, the Malay Archipelago, South America, and the West Indies; while others, such as woad (see Fig. 8), are grown in more temperate European climates. The tinctorial value of these plants was known in India and Egypt long before the Christian era. Egyptian mummy-cloths have been found dyed with indigo. The dye was known to the Greeks and Romans; its use is described by the younger Pliny in his Natural History. Indigo was introduced into Europe about the sixteenth century, but its use was strongly opposed by the woad cultivators, with whose industry the dye came into competition. In France the opposition was strong enough to secure the passing of an act in the time of Henry IV.

COAL; AND WHAT WE GET FROM IT. 125

FIG. 7.—INDIGO PLANT (*Indigofera tinctoria*).

inflicting the penalty of death upon any person found using the dye. The importance of indigo as an article of commerce is sufficiently known at the present time; more than 8000 tons are produced annually, corresponding in money value to about four million pounds. It is of importance to us as rulers of India to remember that the cultivation and manufacture of indigo is one of the staple industries of that country, from which the European markets derive the greater part of their supply.

Imagine the industrial revolution which would be caused by the discovery of a process for obtaining indigo synthetically from a coal-tar hydrocarbon, at a price which would compare favourably with that of the natural product. This has not actually been done as yet, but chemists have attempted to compete with Nature in this direction, and the present state of the competition is that the natural product can be cultivated and made more cheaply. Nevertheless the dye can be synthesised from a coal-tar hydrocarbon, and this is one of the greatest achievements of modern chemistry in connection with the tar-products. For more than half a century indigo had been undergoing investigation by chemists, and at length the work culminated in the discovery of a method for producing it artificially. This discovery was the outcome of the labour of Adolf v. Baeyer, who commenced his researches upon the derivatives of indigo in 1866, and who in 1880 secured the first patents for the

FIG. 8.—WOAD (*Isatis tinctoria*).

manufacture of the colouring-matter. It is to the laborious and brilliant investigations of this chemist that we owe nearly all that is at present known about the chemistry of indigo and allied compounds.

Two methods have been used for the production of artificial indigo—benzal chloride being the starting-point in one of these, and nitrobenzoic aldehyde in the other. The generating hydrocarbon is therefore toluene. By heating benzal chloride with dry sodium acetate there is formed an acid known as cinnamic acid, a fragrant compound which derives its name from cinnamon, because the acid was prepared by the oxidation of oil of cinnamon by Dumas and Peligot in 1834. The acid and its ethers occur also in many balsams, so that we have here another instance of the synthesis of a natural vegetable product from a coal-tar hydrocarbon. The subsequent steps are—(1) the nitration of the acid to produce nitrocinnamic acid ; (2) the addition of bromine to form a dibromide of the nitro-acid ; (3) the action of alkali on the dibromide to produce what is known as "propiolic acid." The latter, under the influence of mild alkaline reducing agents, is transformed into indigo-blue. The process depending on the use of nitrobenzoic aldehyde is much simpler; but the particular nitro-derivative of the aldehyde which is required is at present difficult to make, and therefore expensive. If the production of this compound could be cheapened, the competition between artificial and natural indigo would assume a much more serious aspect.[1]

The light oil of the tar-distiller has now been sufficiently dealt with so far as regards colouring-matters; let us pass on to the next fraction of the tar, the carbolic oil. The important constituents of this portion are carbolic acid and naphthalene. The carbolic oil is in the first place separated into two distinct portions by washing with an alkaline solution. Carbolic acid or phenol belongs to a class of compounds derived from hydrocarbons of the benzene and related series by the substitution of the residue of water for hydrogen. This water-residue is known to chemists as "hydroxyl"—it is water less one atom of hydrogen. Carbolic acid or phenol is hydroxybenzene; and all analogous compounds are spoken of as "phenols." It will be understood in future that a phenol is a hydroxy-derivative of a benzenoid hydrocarbon. Now these phenols are all more or less acid in character by virtue of the hydroxyl-group which they contain. For this reason they dissolve in aqueous alkaline solutions, and are precipitated therefrom by acids.

---

[1] Since the above was written, new synthetical processes for the production of indigo have been made known in Germany by Karl Heumann. Of the commercial aspect of these discoveries it is of course impossible at present to form an opinion.

I

This will enable us to understand the purification of the carbolic oil.

The two layers into which this oil separates after washing with alkali are (1) the aqueous alkaline solution of the carbolic acid and other phenols, and (2) the undissolved naphthalene contaminated with oily hydrocarbons and other impurities. Each of these portions has its industrial history. The alkaline solution, on being drawn off and made acid, yields its mixture of phenols in the form of a dark oil from which carbolic acid is separated by a laborious series of fractional distillations. The undissolved hydrocarbon is similarly purified by fractional distillation, and furnishes the solid crystalline naphthalene. The tar from one ton of Lancashire coal yields about $1\frac{1}{2}$lbs. of carbolic acid, equal to about 1 per cent. by weight of the tar, and about $6\frac{1}{4}$lbs. of naphthalene, so that this last hydrocarbon is one of the chief constituents of the tar, of which it forms from 8 to 10 per cent. by weight.

The crude carbolic acid as separated from the alkaline solution is a mixture of several phenolic compounds, and all of these but the carbolic acid itself are gradually removed during the process of purification. Among the compounds associated with the carbolic acid are certain phenols of higher boiling-point, which bear the same relationship to carbolic acid that toluene bears to benzene. That is to say, that while phenol itself is hydroxybenzene, these other compounds, which are called "cresols," are hydroxytoluenes. The cresols form an oily liquid largely used for disinfecting purposes under the designation of "liquid carbolic acid," or "cresylic acid." Carbolic acid is a white crystalline solid possessing strongly antiseptic properties, and is therefore of immense value in all cases where putrefaction or decay has to be arrested. It was discovered in coal-tar by Runge in 1834, and was obtained pure by Laurent in 1840.

The gradual establishment of the germ-theory of disease, chiefly due to the labours of Pasteur, has led to a most important application of carbolic acid. Once again we find the coal-tar industry brought into contact with another department of science. Arguing from the view that putrefactive change is brought about by the presence of the germs of micro-organisms ever present in the atmosphere, Sir Joseph Lister proposed that during surgical operations the incised part should be kept under a spray of the germicidal carbolic acid to prevent subsequent mortification. No operation upon portions of the body exposed to the air is at present conducted without this precaution, and many a human life must have been saved by Lister's treatment. To this result the chemist and technologist have contributed, not only by the discovery of the carbolic acid in the tar, but also by the development of the necessary processes for its purification. It should be added that the phenol used must be of the greatest possible purity, and the requirements of the surgeon have been met by chemical and technological skill.

From surgery back to colouring-matters, and from these to pharmaceutical preparations and perfumes, are we led in following up the cycles of chemical transformation which these tar-products have undergone in the hands of the technologist, guided by the researches of the chemist. It was observed by Runge in 1834 that crude carbolic acid, on treatment with lime, gave a red, acid colouring-matter which he separated and named " rosolic acid." The observation was followed up, and many other chemists obtained red colouring-matters by the oxidation of crude phenol. In 1859, the colour-giving property of carbolic acid acquired industrial importance from a discovery made by Kolbe and Schmitt in Germany, and by Persoz in France. These chemists found that a good yield of the colouring-matter was obtained by heating phenol with oxalic and sulphuric acids. Under the names of "corallin" and "aurin" the dye-stuff was introduced into commerce, and it is still used for certain purposes, especially for the preparation of coloured lakes for paper-staining.

The scientific development of the history of this phenol dye is full of interest, but we can only give it a passing glance. Its interest lies chiefly in the circumstance that it is related to magenta, as was first pointed out by Caro and Wanklyn in 1866. In fact they obtained rosolic acid from magenta by the action of nitrous acid on the latter. We now know that a diazo-salt is first formed under these circumstances, and that the decomposition of this unstable compound in the presence of water gives rise to the rosolic acid. Later researches have shown that by heating rosolic acid with ammonia it is converted into rosaniline. It is also known that the commercial corallin, like the commercial magenta, is a mixture of closely related colouring-matters. The close analogy between magenta and rosolic acid was further shown by Caro in 1866. In the same way that Hofmann found that magenta could not be produced by the oxidation of *pure* aniline, Caro found that a mixture of phenol and cresol was necessary for the production of rosolic acid when inorganic oxidizers were used. It is indeed this series of investigations upon the phenol dyes—investigations which have been taken part in not only by the chemists named, but also by Graebe, Dale and Schorlemmer, and the Fischers—which led up to the discovery of the constitution of the colouring-matters of the rosaniline group, and, through this, to the far-reaching industrial developments of the discovery as traced in the last chapter. It is evident, from what has been said, that rosolic acid and its related colouring-matters are members of the triphenyl-methane group. They are in fact the hydroxylic or acid analogues of the amido-containing or basic dyes of the rosaniline series.

In the fragrant blossom of the meadowsweet (*Spiræa ulmaria*) there is contained an acid which is found also as an ether in the oil of wintergreen (*Gautheria procumbens*). This is salicylic acid, a white crystalline compound which has been known to chemists since 1839. In 1860 Kolbe prepared the sodium salt of this acid by passing carbon dioxide gas into phenol in which metallic sodium had been dissolved. It was found subsequently that the same transformation was brought about by heating the dry sodium salt of carbolic acid in an atmosphere of carbon dioxide. This process of Kolbe's is now worked on a manufacturing scale for the preparation of artificial salicylic acid. The acid and its salts and ethers find numerous applications as antiseptics, for the preservation of food, and in pharmacy.

Salicylic acid is employed also for the manufacture of certain azo-dyes in a way that it will be very instructive to consider, because the process used may be taken as typical of the general method of preparing such compounds. Solutions of diazo-salts act not only upon amido- and diamido-compounds, as we have seen in the case of aniline yellow and chrysoïdine, but also upon phenols, forming acid azo-colours. This important fact was made known in 1870 by the German chemists Kekulé and Hidegh, but more than six years elapsed before this discovery was taken advantage of by the technologist. Large numbers of these acid azo-dyes are now made from various diazotised amido-compounds combined with different phenols and phenolic acids. The mode of procedure is to diazotise the amido-compound by sodium nitrite and hydrochloric acid in the manner already described, and then add the diazo-salt solution to the phenolic compound dissolved in alkali. The colouring-matter is at once formed. Salicylic acid possesses the characters both of an acid and a phenol. It combines readily with diazo-salts under the circumstances described, and gives rise to azo-dyes, some of which are of technical value.

The manufacture of azo-dyes from salicylic acid brings us into contact with certain amidic compounds which figure so largely in the tinctorial industry that they may be conveniently dealt with here. These bases are not azo-compounds themselves, but they are prepared from azo-compounds, viz. from the azobenzene and azotoluene which were spoken about in the last chapter. When these are reduced by acid reducing-agents, they become converted into diamido-bases which are known as benzidine and tolidine respectively. These bases can be diazotised, and as they contain two amido-groups, they form double diazo-salts, *i. e.* tetrazo-salts, which are capable of combining with amido-compounds, or phenols, in the usual way. Thus diazotised benzidine and tolidine combine with salicylic acid to form valuable yellow azo-dyes known as "chrysamines." The dyes of this class obviously contain two azo-groups.

Some other uses of carbolic acid must next be considered. Of the colouring-matters derived from coal-tar, none is more widely known than the oldest artificial yellow dye, picric acid. This is a phenol derivative, and was first obtained as long ago as 1771 by Woulfe, by acting upon indigo with nitric acid. Laurent in 1842 was the first to obtain this dye from carbolic acid, from which compound it is still manufactured by acting upon the sulpho-acid with nitric acid. Chemically considered, it is trinitrophenol. It has a very wide application as a dye, and has been used as an explosive agent. A similar colouring-matter was made from cresol in 1869, and introduced under the name of "Victoria yellow," which is dinitrocresol. Other dyes derived directly or indirectly from phenol will take us back once again to toluene.

A new diazotisable diamido-compound was obtained from this last hydrocarbon, and introduced in 1886 by Leonhardt & Co. One of the three isomeric nitrotoluenes furnishes a sulpho-acid which, on treatment with alkali, gives a compound derived from a hydrocarbon known as stilbene, and this, on reduction, is converted into the diamido-compound referred to. The latter, which is a disulpho-acid as well as a diamido-compound, can be diazotised and combined with phenols, &c. The stilbene azo-dyes thus prepared from phenol and salicylic acid, like the chrysamines, are yellow colouring-matters, containing two azo-groups. It is a valuable characteristic of these secondary azo-dyes that they all possess a special affinity for vegetable fibre, and their introduction has exerted a great influence upon the art of cotton-dyeing. We shall have to return to these cotton-dyes again shortly.

Before leaving this branch of the subject, the following scheme is presented to show the relationships and inter-relationships of the products thus far dealt with in the present chapter—

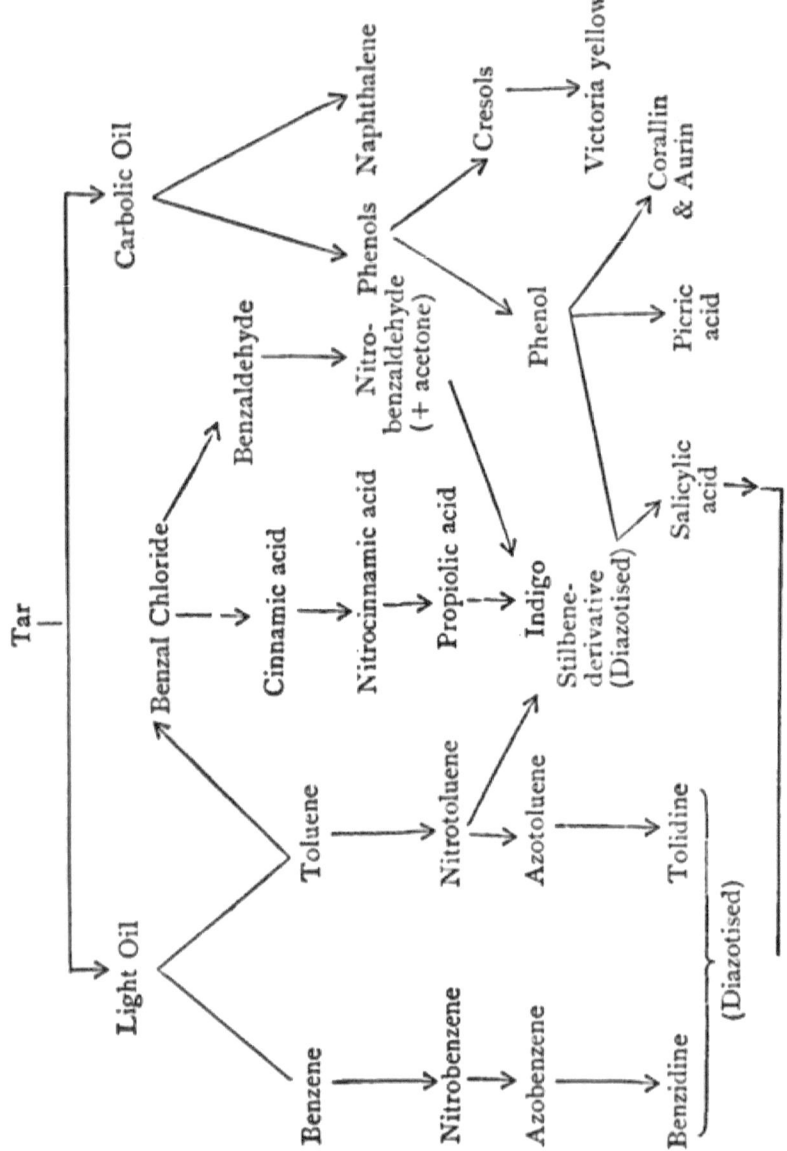

The existence of naphthalene in coal-tar was made known in 1820 by Garden, who gave it this name because the oils obtained from the tar by distillation went under the general designation of naphtha. The greater portion of the hydrocarbon is contained in the carbolic oil, and is separated and purified in the manner described. A further quantity of impure naphthalene separates out from the next fraction—the creosote oil, and this is similarly washed and purified by distillation. The large quantity of naphthalene existing in tar has already been referred to, but although it is such an important constituent, it was only late in the history of the colour industry that it found any extensive application. In early times it was regarded as a nuisance, and was burnt as fuel, or for the production of a dense soot, which was condensed to form lampblack. It will be remembered that the first of the coal-tar colours made required only the light oils. There are at present only a few direct uses for naphthalene, but one of its applications is sufficiently important to be mentioned.

The hydrocarbon is a white crystalline solid melting at 80° C., and boiling at 217° C. Although it has a high boiling-point, it passes readily into vapour at lower temperatures, and the vapour on condensation forms beautiful silvery crystalline scales. This product is "sublimed naphthalene."

The vapour of naphthalene burns with a highly luminous flame, and if mixed with coal-gas, it considerably increases the luminosity of the flame. Advantage is taken of this in the so-called "albo-carbon light," which is the flame of burning coal-

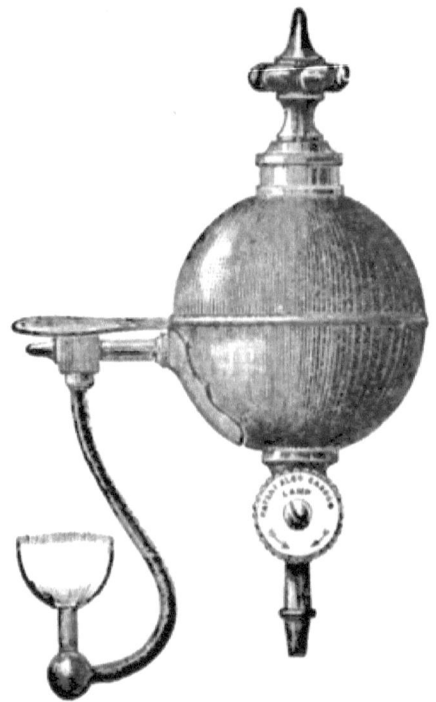

FIG. 9.—ALBO-CARBON BURNER.

gas saturated with naphthalene vapour. The burner is constructed so that the gas passes through a reservoir filled with melted naphthalene kept hot by the flame itself (Fig. 9).

To appreciate properly the value of those discoveries which have enabled manufacturers to utilize this hydrocarbon, it is only necessary to recall to mind the actual quantity produced in this country. Supposing that ten million tons of coal are used annually for gas-making, and that the 500,000 tons of tar resulting therefrom contain only eight per cent. of naphthalene, there would be available about 40,000 tons of this hydrocarbon annually. Great as have been the recent advancements in the utilization of naphthalene derivatives, there is still a larger quantity of this hydrocarbon produced than is necessary to supply the wants of the colour-manufacturer. From this last statement it will be inferred that naphthalene is now a source of colouring-matters. Let us consider how this has been brought about.

The phenols of naphthalene are called naphthols —they bear the same relationship to naphthalene that carbolic acid bears to benzene. Owing to the structure of the naphthalene molecule there are two isomeric naphthols, whereas there is only one phenol. The naphthols—known as alpha- and beta-naphthol, respectively—are now made on a large scale from naphthalene, by heating the latter with sulphuric acid; at a low temperature the alpha sulpho-acid is produced, and at a higher temperature the beta sulpho-acid, and these acids on fusion with caustic soda furnish the corresponding naphthols. Similarly there are two amidonaphthalenes, known as alpha- and beta-naphthylamine respectively. As aniline is to benzene, so are the naphthylamines to naphthalene. The alpha-compound is made in precisely the same way as aniline, viz. by acting upon naphthalene with nitric acid so as to form nitronaphthalene, and then reducing the latter with iron dust and acid. Beta-napthylamine cannot be made in this way; it is prepared from beta-naphthol by heating the latter in presence of ammonia, when the hydroxyl becomes replaced by the amido-group in accordance with a process patented in 1880 by the Baden Aniline Company. The principle thus utilized is the outcome of the scientific work of two Austrian chemists, Merz and Weith. Setting out from the naphthols and naphthylamines we shall be led into industrial developments of the greatest importance.

The first naphthalene colour was a yellow dye, discovered by Martius in 1864, and manufactured under the name of "Manchester yellow." It is, chemically speaking, dinitro-alpha-naphthol; but it was not at first made from naphthol, as the latter was not at the time a technical product. It was made from alpha-naphthylamine by the action of nitrous and nitric acids. When a good method for making the naphthol was discovered in 1869, the dye was made from this. The process is just the same as that employed in making picric acid; the naphthol is converted into a sulpho-acid, and this when acted upon by nitric acid, gives the colouring-matter. Manchester yellow is now largely used for colouring soap, but as a dye-stuff it has been improved upon in a manner that will be readily understood. The original colouring-matter being somewhat fugitive, it was found that its sulpho-acid was much faster. This sulpho-acid cannot be made by the direct action of sulphuric acid upon the colouring-matter—as in the case of acid yellow or acid magenta—but by acting upon the naphthol with very strong sulphuric acid, three sulphuric acid residues or sulpho-groups enter the molecule, and then on nitration only two of these are replaced by nitro-groups, and there results a sulpho-acid of dinitro-alpha-naphthol. This was discovered in 1879 by Caro, and introduced as "acid naphthol yellow." It is now one of the standard yellow dyes.

The history of another important group of colouring-matters dependent on naphthalene begins with A. v. Baeyer in 1871 and with Caro in 1874. Two products formerly known only as laboratory preparations were called into requisition by this discovery. One of these compounds, phthalic acid, is obtained from naphthalene, and the other, resorcin or resorcinol, is prepared from benzene. Phthalic acid, which was discovered in 1836 by Laurent, is a product of the oxidation of many benzenoid compounds. Chemically considered it is a di-derivative of benzene, *i. e.* two of the hydrogen atoms of benzene are replaced by certain groups of carbon, oxygen, and hydrogen atoms. We have seen how the replacement of hydrogen by an ammonia-residue, amidogen, gives rise to bases such as amidobenzene (aniline), or diamidobenzene. Similarly, the replacement of hydrogen by a water-residue, hydroxyl, gives rise to a phenol. The group of carbon, oxygen, and hydrogen atoms which confers the property of acidity upon an organic compound is a half-molecule of oxalic acid —it is known as the carboxyl group. Thus benzoic acid is the carboxyl-derivative of benzene, and the phthalic acid with which we are now concerned is a dicarboxyl-derivative of benzene. It is related to benzoic acid in the same way that diamido-benzene is related to aniline. Three isomeric phthalic acids are known, but only one of these is of use in the present branch of manufacture. The acid in question, although a derivative of benzene, is most economically prepared by the oxidation of certain derivatives of naphthalene which, when completely broken down by energetic oxidizing agents, furnish the acid. Thus the dinitronaphthol described as Manchester yellow, if heated for some time with dilute nitric acid, furnishes phthalic acid.

The latter is made on a large scale by the oxidation of a compound which naphthalene forms with chlorine, and known as naphthalene tetrachloride, because it contains four atoms of chlorine.

The other compound, resorcinol, was known to chemistry ten years before it was utilized as a source of colouring-matters. It was originally prepared by fusing certain resins, such as galbanum, asafœtida, &c., with caustic alkali. Soon after its discovery, viz. in 1866, it was shown by Körner to be a derivative of benzene, and from this hint the technical process for the preparation of the compound on a large scale has been developed. Resorcinol is a phenolic derivative of benzene containing two hydroxyl groups; it is therefore related to phenol in the same way that diamidobenzene is related to aniline or phthalic acid to benzoic acid. The relationships can be expressed in a tabular form thus—

Amidobenzene or Aniline.	Benzoic acid.	Carbolic acid or Phenol.
Diamidobenzene.	Phthalic acid.	Resorcinol.

Resorcinol is now made by heating benzene with very strong sulphuric acid so as to convert it into a disulpho-acid, and the sodium salt of the latter is then fused with alkali. As a technical operation it is one of great delicacy and skill, and the manufacture is confined to a few Continental factories.

When phthalic acid is heated it loses water, and is transformed into a white, magnificently crystalline substance known as phthalic anhydride, *i.e.* the acid deprived of water. In 1871, A. v. Baeyer, the eminent chemist who subsequently synthesised indigo, published the first of a series of investigations describing the compounds produced by heating phthalic anhydride with phenols. To these compounds he gave the name of "phthaleïns." Baeyer's work, like that of so many other chemists who have contributed to the advancement of the coal-tar colour industry, was of a purely scientific character at first, but it soon led to technological developments. The phthaleïns are all acid compounds possessing more or less tinctorial power. One of the first discovered was produced by heating phthalic anhydride with an acid known as gallic acid, which occurs in vegetable galls, and in the form of tannin in many vegetable extracts which are used by the tanner. The acid is a phenolic derivative of benzoic acid, viz. trihydroxybenzoic acid, and on heating it readily passes into trihydroxybenzene, which is the "pyrogallic acid" or pyrogallol familiar as a photographic developer. The phthaleïn formed from gallic acid and phthalic anhydride really results from the union of the latter with pyrogallol. It is now manufactured under the name of "galleïn," and is largely used for imparting a bluish grey shade to cotton fabrics. By heating galleïn with strong sulphuric acid, it is transformed into another colouring-matter which gives remarkably fast olive-green shades when dyed on cotton fibre with a suitable mordant. This derivative of galleïn is used to a considerable extent under the name of "cœruleïn." These two colouring-matters were the first practical outcome of v. Baeyer's researches. There is another possible development in this direction which chemistry may yet accomplish, and another natural colouring-matter may be threatened, even as the indigo culture was threatened by the later work of the same chemist. There is reason for believing that the colouring-matter of logwood, known to chemists as hæmateïn, is related to or derived in some way from the phthaleïns, and the synthesis of this compound may ultimately be effected.

The dye introduced by Caro in 1874 is the brominated phthaleïn of resorcinol. The phthaleïn itself is a yellow dye, and the solutions of its salts show a splendid and most intense greenish yellow fluorescence, for which reason it is called "fluoresceïn." When brominated, the latter furnishes a beautiful red colouring-matter known as "eosin" (Gr. ἔως, dawn), and the introduction of this gave an industrial impetus to the phthaleïns which led to the discovery of many other related colouring-matters now largely used under various trade designations. About a dozen distinct compounds producing different shades of pink, crimson and red, and all derived from fluoresceïn, are at present in the market, and a few other phthaleïns formed by heating phthalic anhydride with other phenolic compounds instead of resorcinol or pyrogallol (*e. g.* diethylamidophenol), are also of industrial importance. By converting nitrobenzene into a sulpho-acid, reducing to an amido-sulpho-acid, and then fusing with alkali, an amido-phenol is produced, the ethers of which, when heated with phthalic anhydride, give rise to red phthaleïns of most intense colouring power introduced by the Baden Aniline Company as "rhodamines."

It remains to point out that the scientific spirit which prompted the investigation of the phthaleïns in the first instance has followed these compounds throughout their technological career. The researches started by v. Baeyer were taken up by various chemists, whose work together with that of the original discoverer has led to the elucidation of the constitution of these colouring-matters. The phthaleïns are members of the triphenyl-methane group, and are therefore related to magenta, corallin, malachite green, methyl violet, and the phosgene dyes.

It has been said that the phenolic and amidic

derivatives of naphthalene, *i.e.* the naphthols and naphthylamines, are of the greatest importance to the colour industry. One of the first uses of alpha-naphthylamine has already been mentioned, viz. for the production of the Manchester yellow, which was afterwards made more advantageously from alpha-naphthol. A red colouring-matter possessing a beautiful fluorescence was afterwards (1869) made from this naphthylamine and introduced as "Magdala red." The latter was discovered by Schiendl of Vienna in 1867. It was prepared in precisely the same way as induline was prepared from aniline yellow. The latter, which is amido-azo-benzene, and which is prepared, broadly speaking, by the action of nitrous acid on aniline, has its analogue in amido-azonaphthalene, which is similarly prepared by the action of nitrous acid on naphthylamine. Just as aniline yellow when heated with aniline and an aniline salt gives induline, so amido-azonaphthalene when heated with naphthylamine and a salt of this base gives Magdala red. The latter is, therefore, a naphthalene analogue of induline, as was shown by Hofmann in 1869, and the knowledge of the constitution of the azines which has been gained of late years, enables us to relegate the colouring-matter to this group. This knowledge has also enabled the manufacture to be conducted on more rational principles, viz. by the method employed for the production of the saffranines, as previously sketched.

The introduction of azo-dyes, formed by the action of a diazotised amido-compound on a phenol or another amido-compound, marks the period from which the naphthols and naphthylamines rose to the first rank of importance as raw materials for the colour manufacturer. The introduction of chrysoïdine in 1876 was immediately followed by the manufacture of acid azo-dyes obtained by combining diazotised amido-sulpho-acids with phenols of various kinds, or with bases, such as dimethylaniline and diphenylamine. From what has been said in the foregoing portion of this volume, it is evident that all such azo-compounds result from the combination of two things, viz. (1) a diazotised amido-compound, and (2) a phenolic or amidic compound. Either (1) or (2) or both may be a sulpho-acid, and the resulting dye will then also be a sulpho-acid.

The first of these colouring-matters derived from the naphthols, was introduced in 1876-77 by Roussin and Poirrier, and by O. N. Witt. They were prepared by converting aniline into a sulpho-acid (sulphanilic acid), diazotising this and combining the diazo-compound with alpha- or beta-naphthol. The compounds formed are brilliant orange dyes, the latter being still largely consumed as "naphthol orange." Other dye-stuffs of a similar nature were introduced by Caro about the same time, and were prepared from the diazotised sulpho-acid of alpha-naphthylamine combined with the naphthols. By this means alpha-naphthol gives what is known as "acid brown," or "fast brown," and beta-naphthol a fine crimson, known as "fast red," or "roccellin." Diazotised compounds combine also with this same sulpho-acid of alpha-naphthylamine (known as naphthionic acid), the first colouring-matter formed in this way having been introduced by Roussin and Poirrier in 1878. It was prepared by diazotising a nitro-derivative of aniline, and acting with the diazo-salt on napthionic acid, and this dye is still used to some extent under the name of "archil substitute." In 1878, the firm of Meister, Lucius & Brüning of Höchst-on-the-Main gave a further impetus to the utilization of naphthalene by discovering two isomeric disulpho-acids of beta-naphthol formed by heating that phenol with sulphuric acid. By combining various diazotised bases with these sulpho-acids, a splendid series of acid azo-dyes ranging in shade from bright orange to claret-red, and to scarlets rivalling cochineal in brilliancy were given to the tinctorial industry.

The colouring-matters introduced in 1878 by the Höchst factory under the names of "Ponceaux" of various brands, and "Bordeaux," although to some extent superseded by later discoveries, still occupy an important position. Their discovery not only increased the consumption of beta-naphthol, but also that of the bases which were used for diazotising. These bases are alpha-naphthylamine and those of the aniline series. The intimate relationship which exists between chemical science and technology—a relationship which appears so constantly in the foregoing portions of this work— is well brought out by the discovery under consideration. A little more chemistry will enable this statement to be appreciated.

Going back for a moment to the hydrocarbons obtained from light oil, it will be remembered that benzene and toluene have thus far been considered as the only ones of importance to the colourmaker. Until the discovery embodied in the patent specification of 1878, the portions of the light oil boiling above toluene were of no value in the colour industry. Benzene and toluene are related to each other in a way which chemists describe by saying that they are "homologous." This means that they are members of a regularly graduated series, the successive terms of which differ by the same number of atoms of carbon and hydrogen. Thus toluene contains one atom of carbon and two atoms of hydrogen more than benzene. Above toluene are higher homologues, viz. xylene, cumene, &c , which occur in the light oil, the former being related to toluene in the same way that toluene is related to benzene, while cumene again contains one atom of carbon and two atoms of hydrogen more than xylene. This relationship between the members of homologous series is expressed in other terms by saying that the weight of the molecule increases by a constant quantity as we ascend the series.

The homology existing among the hydrocarbons extends to all their derivatives. Thus phenol is the lower homologue of the cresols. Also we have the homologous series—

Benzene.	Nitrobenzene.	Aniline.
Toluene.	Nitrotoluene.	Toluidine.
Xylene.	Nitroxylene.	Xylidine.
Cumene.	Nitrocumene.	Cumidine.

The bases of the third column when diazotised and combined with the disulpho-acids of beta-naphthol give a graduated series of dyes beginning with orange and ending with bluish scarlet. Thus it was observed that the toluidine colour was redder than the aniline colour, and it was a natural inference that the xylidine colour would be still redder. At the time of this discovery no azo-colour of a true scarlet shade had been manufactured successfully. A demand for the higher homologues of benzene was thus created, and the higher boiling-point fractions of the light oil, which had been formerly used as solvent naphtha, became of value as sources of colouring-matters. The isolation of coal-tar xylene (which is a mixture of three isomeric hydrocarbons) is easily effected by fractional distillation with a rectifying column, and by nitration and reduction, in the same way as in the manufacture of aniline, xylidine is placed at the disposal of the colour-maker.

Xylidine scarlet, although at the time of its introduction the only true azo-scarlet likely to come into competition with cochineal, was still somewhat on the orange side. The cumidine dye would obviously be nearer the desired shade. To meet this want, cumidine had to be made on a large scale, but here practical difficulties interposed themselves. The quantity of cumene in the light oil is but small, and it is associated with other hydrocarbons which are impurities from the present point of view, and from which it is separated only with difficulty. A new source of the base had therefore to be sought, and here again we find chemical science ministering to the wants of the technologist.

It was explained in the last chapter that aniline and similar bases can be methylated by heating their dry salts with methyl alcohol under pressure,

In this way dimethylaniline is made, and dimethyltoluidine or dimethylxylidine can similarly be prepared. Now it was shown by Hofmann in 1871, that if this operation is conducted at a very high temperature, and under very great pressure, the methyl-alcohol residue, *i.e.* the methyl-group, does not replace the amidic hydrogen or the hydrogen of the ammonia residue, but the methylation takes place in another way, resulting in the formation of a higher homologue of the base started with. For example, by heating aniline salt and pure wood-spirit to a temperature considerably above that necessary for producing dimethylaniline, toluidine is formed. In a similar way, by heating xylidine hydrochloride and methyl alcohol for some time in a closed vessel at about 300° C. cumidine is produced. Hofmann's discovery was thus utilized in 1882, and by its means the base was manufactured, and cumidine scarlet, very similar in shade to cochineal, became an article of commerce.

While the development of this branch of the colour industry was taking place by means of the new naphthol disulpho-acids, the cultivation of the fertile field of the azo-dyes was being carried on in other directions. It came to be realized that the fundamental discovery of Griess was capable of being extended to all kinds of amido-compounds.

The azo-dyes hitherto introduced had all been derived from amido-compounds containing only one amido-group, and they accordingly contained only one azo-group; they were *primary* azo-compounds. It was soon found that aniline yellow, which already contains one azo-group as well as an amido-group, could be again diazotised and combined with phenols so as to produce compounds containing two azo-groups, *i.e. secondary* azo-compounds. The sulpho-acid of aniline yellow —Grässler's "acid yellow"—was the first source of azo-dyes of this class. By diazotising this amidoazo-sulpho-acid, and combining it with beta-naphthol, a fine scarlet dye was discovered by Nietzki in 1879, and introduced under the name of "Biebrich scarlet." Two years later a new sulpho-acid of beta-naphthol was discovered by Bayer & Co. of Elberfeld, and this gave rise, when combined with diazotised acid yellow and analogous compounds, to another series of brilliant dyes introduced as "Crocein scarlets."

From these beginnings the development of the azo-dyes has been steadily carried on to the present time—year by year new diazotisable amido-compounds or new sulpho-acids of the naphthols and naphthylamines are being discovered, and this branch of the colour industry has already assumed colossal dimensions. An important departure was made in 1884 by Böttiger, who introduced the first secondary azo-colours derived from benzidine. As already explained in connection with salicylic acid, this base and its homologue tolidine form tetrazo-salts, which combine with phenols and amines or their sulpho-acids. One of the first colouring-matters of this group was obtained by combining diazotised benzidine with the sulpho-acid of alpha-naphthylamine (naphthionic acid), and was introduced under the name of "Congo red." Then came the discovery (Pfaff, 1885), that the tetrazo-salts of benzidine and tolidine combine with phenols, amines, &c., in two stages, one of the diazo-groups first combining with one-half of the whole quantity of phenol to form an intermediate compound, which then combines with the other half of the phenol to form the secondary azo-dye. In the hands of the "Actiengesellschaft für Anilin-fabrikation" of Berlin this discovery has been utilized for the production of a number of such azo-colours containing two distinct phenols, or amines, or sulpho-acids. Tolidine has been found to give better colouring-matters in most cases than benzidine, and it is scarcely necessary to point out that an increased demand for the nitrotoluene from which this base is made is the necessary consequence of this discovery.

It is impossible to attempt to specify by name

any of these recent benzidine and tolidine dyes. Their introduction has been the means of finding new uses for the naphthylamines and naphthols and their sulpho-acids, and has thus contributed largely to the utilization of naphthalene. An impetus has been given to the investigation of these sulpho-acids, and chemical science has profited largely thereby. The process by which beta-naphthylamine is prepared from beta-naphthol, already referred to, viz. by heating with ammonia under pressure, has been extended to the sulpho-acids of beta-naphthol, and by this means new beta-naphthylamine sulpho-acids have been prepared, and figure largely in the production of these secondary azo-colours. The latter, as previously stated, possess the most valuable property of dyeing cotton fibre directly, and by their means the art of cotton dyeing has been greatly simplified. The shades given by these colours vary from yellow through orange to bright scarlet, violet, or purple.

In addition to benzidine and tolidine, other diazotisable amido-compounds have of late years been pressed into the service of the colour-manufacturer. The derivative of stilbene, already mentioned as being prepared from a sulpho-acid of one of the nitrotoluenes, forms tetrazo-salts, which can be combined with similar or dissimilar phenols, amines, or sulpho-acids, as in the case of benzidine and tolidine. Various shades of red and purple are thus obtained from the diazotised compound, when the latter is combined with the naphthylamines, naphthols, or their sulpho-acids. These, again, are all cotton dyes. The nitro-derivatives of the ethers of phenol and cresol, when reduced in the same way that nitrobenzene and nitrotoluene are reduced to azobenzene and azotoluene, also furnish azo-compounds which, on further reduction, give bases analogous to benzidine and tolidine. Secondary azo-colours derived from these bases and the usual naphthalene derivatives are also manufactured. It is among the secondary azo-dyes that we meet with the first direct dyeing blacks, the importance of which will be realized when it is remembered that the ordinary aniline-black is not adapted for wool dyeing. The azo-blacks are obtained by combining diazotised sulpho-acids of amidoazocompounds of the benzene or naphthalene series with naphthol sulpho-acids or other naphthalene derivatives.

One other series of azo-compounds must be briefly referred to. It has long been known that aniline and toluidine when heated with sulphur evolve sulphuretted hydrogen and give rise to thiobases, that is, aniline or toluidine in which the hydrogen is partly replaced by sulphur. One of the toluidines treated in this way is transformed into a thiotoluidine which, when diazotised and combined with one of the disulpho-acids of beta-naphthol, forms a red azo-dye, introduced by Dahl & Co. in 1885 as "thiorubin." By modifying the conditions of reaction between the sulphur and the base, it was found in 1887 by Arthur Green, that a complicated thio-derivative of toluidine could be produced which possessed very remarkable properties. The sulpho-acid of the thio-base is a yellow dye, which was named by its discoverer "primuline." Not only is primuline a dye, but it contains an amido-group which can be diazotised. If therefore the fabric dyed with primuline is passed through a nitrite bath, a diazo-salt is formed in the fibre, and on immersing the latter in a second bath containing naphthol or other phenol or an amine, an azo-dye is precipitated in the fibre. By this means there are produced valuable "ingrain colours" of various shades of red, orange, purple, &c.

So much for the azo-dyes, one of the most prolific fields of industrial enterprise connected with coal-tar technology. From the introduction of aniline yellow in 1863 to the present time, about 150 distinct compounds of this group have been given to the tinctorial industry. Of these over thirty are cotton dyes containing two azo-groups. Sombre shades, rivalling logwood black, bright yellows, orange-reds, browns, violets, and brilliant scarlets equalling cochineal, have been evolved from the refuse of the gas-works. The artificial colouring-matters have in this last case once again threatened a natural product, and with greater success than the indigo synthesis, for the introduction of the azo-scarlets has caused a marked decline in the cochineal culture.

In addition to the azo-colours, there are certain other products which claim naphthalene as a raw material. In 1879 it was found that one of the sulpho-acids of beta-naphthol when treated with nitrous acid readily gave a nitroso-sulpho-acid. A salt of this last acid, containing sodium and iron as metallic bases, was introduced in 1884, under the name of "naphthol green." It is used both as a dye for wool and as a pigment. It may be mentioned here that other nitroso-derivatives of phenols, such as those of resorcinol and the naphthols, under the name of "gambines," are largely used for dyeing purposes, owing to the facility with which they combine with metallic mordants to form coloured salts in the fibre. In this same year, 1879, it was found that by heating nitrosodimethylaniline with beta-naphthol in an appropriate solvent, a violet colouring-matter was formed. This is now manufactured under the name of "new blue," or other designations, and is largely used for producing an indigo-blue shade on cotton prepared with a suitable mordant. The discovery of this colouring-matter gave an impetus to further discoveries in the same direction. It was found that nitrosodimethylaniline reacted in a similar way with other phenolic or with amidic compounds. In 1881 Köchlin introduced an analogous dye-stuff prepared by the action of the same nitroso-compound on gallic acid. Gallocyanin, as it is called, imparts a violet blue shade to mordanted cotton. Other colouring-matters of the same group are in use; some of them, like "new blue," being derivatives of naphthalene. These compounds all belong to a series of which the parent substance is constructed on a type similar to azine; it contains a nitrogen and oxygen atom linking together the hydrocarbon residues, and is therefore known as "oxazine." The researches of Nietzki in 1888 first established the true constitution of the oxazines.

Closely related to this group is a colouring-matter introduced by Köchlin and Witt in 1881 under the name of "indophenol." It is prepared in the same way as the azines of the "neutral red" group; viz. by the action of nitrosodimethylaniline on alpha-naphthol, or by oxidizing amidodimethylaniline in the presence of alpha-naphthol. Indophenol belongs to that group of blue compounds formed as intermediate products in the manufacture of azines, as mentioned in connection with "neutral red." But while these intermediate blues resulting from the oxidation of a diamine in the presence of another amine are unstable, and pass readily into red azines, indophenol is stable, and can be used for dyeing and printing in the same way as indigo. The shades which it produces are very similar to this last dye, but for certain practical reasons it has not been able to compete with the natural dye-stuff.

The story of naphthalene is summarized in the schemes on pp. 164, 165.

The fraction of coal-tar succeeding the carbolic oil, viz. the creosote oil, does not at present supply the colour manufacturer with any raw materials beyond the small proportion of naphthalene which separates from it in a very impure condition as "creosote salts." This oil consists of a mixture of the higher homologues of phenol with various hydrocarbons and basic compounds. It is the oil used for creosoting timber in the manner already described; and among its other applications may be mentioned its use as an illuminating agent and as a source of lampblack. In order to burn the oil effectively as a source of light, a specially-constructed burner is used, which is fed by a stream of oil raised from a reservoir at its foot by means of compressed air, which also aids the combustion of the oil. There is produced by this means a great

164 COAL;

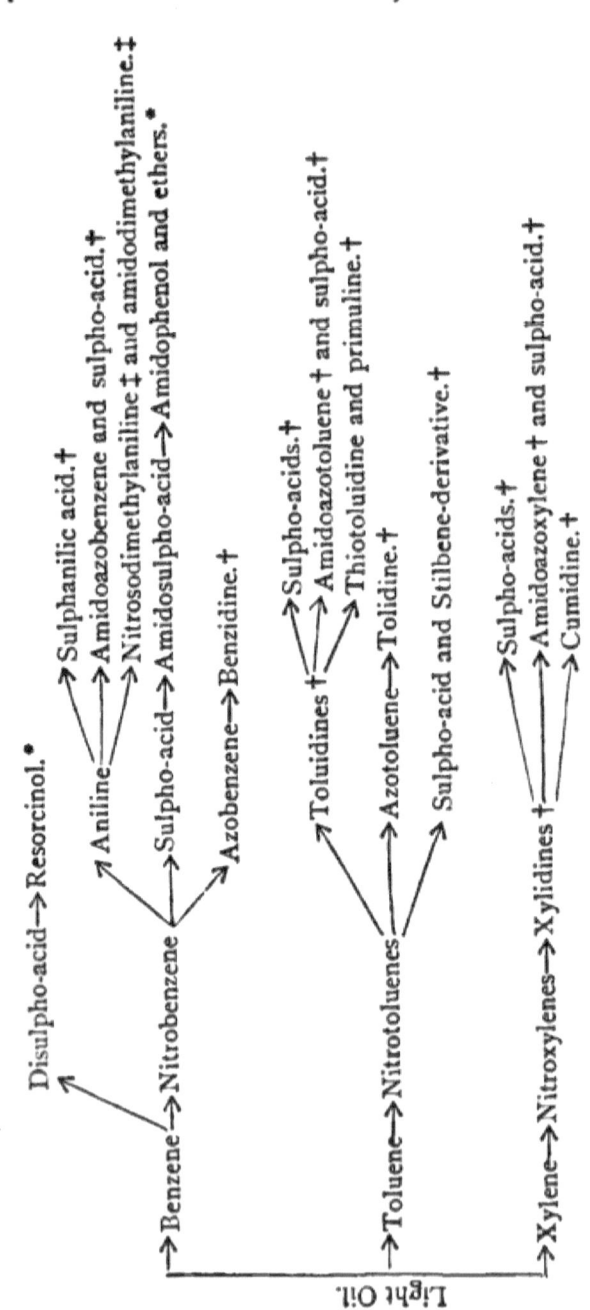

* With phthalic anhydride gives fluorescein and dyes of eosin and rhodamin series.
† Diazotised and combined with naphthylamines, naphthols, or their sulpho-acids give azo-dyes.
‡ With naphthols give oxazines and indophenol.

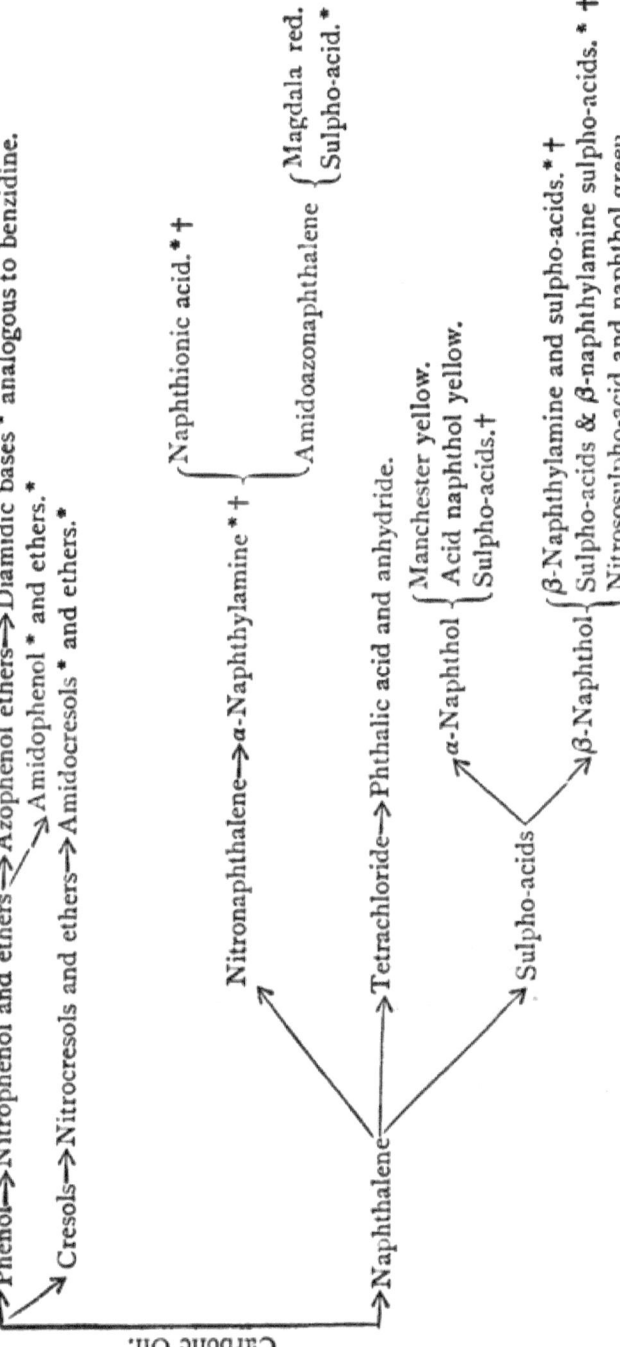

body of lurid flame, which is very serviceable where building or other operations have to be carried on at night (see Fig. 10). For lampblack the oil is simply burnt in iron pans set in ovens, and the sooty smoke conducted into condensing chambers. The creosote oil constitutes more than 30 per cent. by weight of the tar —the time may come when this fraction, like the light oil and carbolic oil, may be found to contain compounds of value to the colour-maker or to other branches of chemical manufacture.

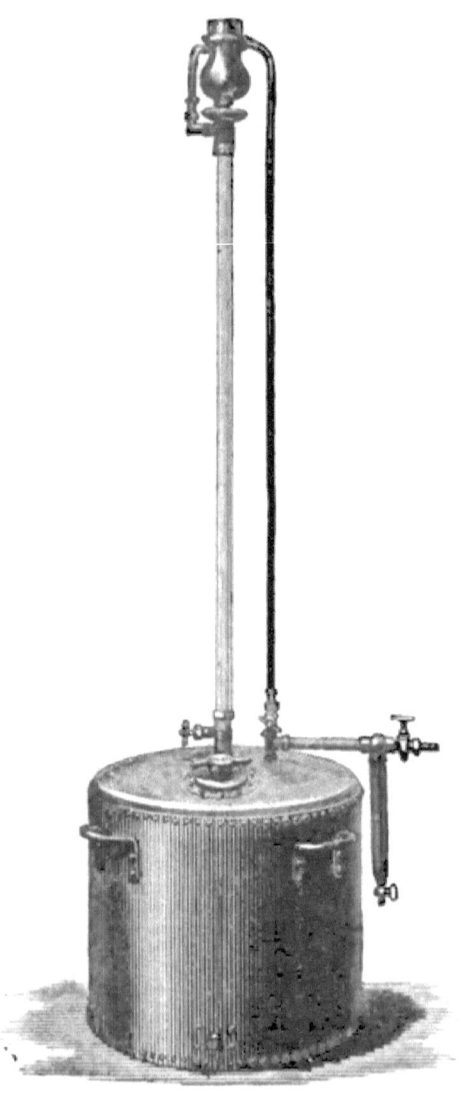

FIG. 10.—VERTICAL BURNER FOR HEAVY COAL OIL BY THE LUCIGEN LIGHT CO.

The utilization

of the next fraction, anthracene oil, is one of the greatest triumphs which applied chemical science

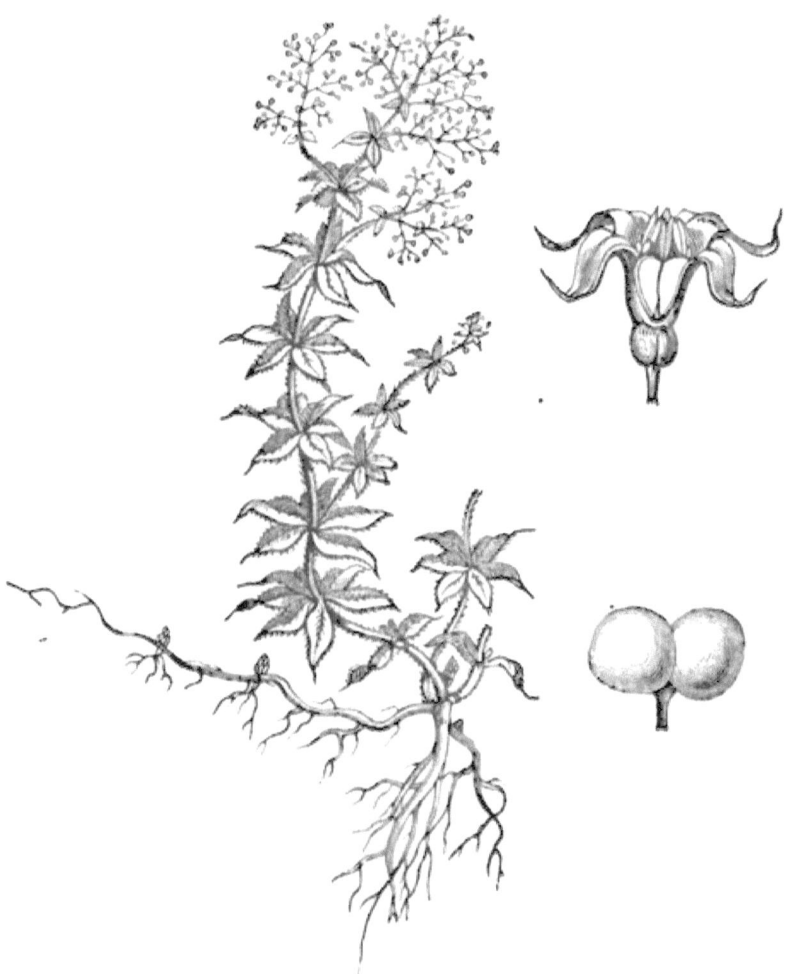

FIG. 11.—THE MADDER PLANT (*Rubia tinctoria*).

can lay claim to since the foundation of the coaltar colour industry. This discovery dates from 1868, when it was shown by two German chemists, Graebe and Liebermann, that the colouring-matter of madder was derived from the hydrocarbon anthracene. Like indigo, madder may be regarded as one of the most ancient of natural dye-stuffs. It consists of the powdered roots of certain plants of the genus *Rubia*, such as *R. tinctoria* (see Fig. 11), *R. peregrina*, and *R. munjista*, which were at one time cultivated on an enormous scale in various parts of Europe and Asia. It is estimated that at the time of Graebe and Liebermann's discovery, 70,000 tons of madder were produced annually in the madder-growing countries of the world. At that time we were importing madder into this country at the rate of 15,000 to 16,000 tons per annum, at a cost of £50 per ton. In ten years the importation had fallen to about 1600 tons, and the price to £18 per ton. At the present time the cultivation of madder is practically extinct.

There is no better gauge of the practical utility of a scientific discovery than the financial effect. In addition to madder, a more concentrated extract containing the colouring-matter itself was largely used by dyers and cotton printers under the name of "garancin." In 1868 we were importing, in addition to the 15,000 to 16,000 tons of madder, about 2000 tons of this extract annually, at a cost of £150 per ton. By 1878 the importation of garancin had sunk to about 140 tons, and the price had been lowered to £65 per ton. The total value of the imports of madder and garancin in 1868 was over one million pounds sterling; in ten years the value of these same imports had been reduced to about £38,000. Concurrently with this falling off in the demand for the natural colouring-matter, the cultivation of the madder plant had to be abandoned, and the vast tracts of land devoted to this purpose became available for other crops. A change amounting to a revolution was produced in an agricultural industry by a discovery in chemistry.

In the persons of two Frenchmen, Messrs. Robiquet and Colin, science laid hands on the colouring-matter of the *Rubia* in 1826. These chemists isolated two compounds which they named alizarin and purpurin. It is now known that there are at least six distinct colouring-matters in the madder root, all of these being anthracene derivatives. It is known also that the colouring-matters do not exist in the free state in the plant, but in the form of compounds known as glucosides, *i.e.* compounds consisting of the colouring-matter combined with the sugar known as glucose. It may be mentioned incidentally that the colouring-matter of the indigo plant also exists as a glucoside in the plant. During a period of more than forty years from the date of its isolation, alizarin was from time to time submitted to examination by chemists, but its composition was not completely established till 1868, when Graebe and Liebermann, by heating it with zinc-dust, obtained anthracene. This was the discovery which gave the death-blow to the madder culture, and converted the last fraction of the tar-oil from a waste product into a material of the greatest value.

The large quantity of madder consumed for tinctorial purposes is indicative of the value of this dye-stuff. It produces shades of red, purple, violet, black, or deep brown, according to the mordant with which the fabric is impregnated. The colours obtained by the use of madder are among the fastest of dyes, the brilliant "Turkey red" being one of the most familiar shades. The discovery of the parent hydrocarbon of this colouring-matter which had been in use for so many ages—a colouring-matter capable of furnishing both in dyeing and printing many distinct shades, all possessed of great fastness—was obviously a step towards the realization of an industrial triumph, viz. the chemical synthesis of alizarin. Within a year of their original observation, this had been accomplished by Graebe and Liebermann, and almost simultaneously by W. H. Perkin in this country. From that time the anthracene, which had previously been burnt or used as lubricating grease, rose in value to an extraordinary extent. In two years a material which could have been bought for a few shillings the ton, rose at the touch of chemical magic to more than two hundred times its former value.

Anthracene is a white crystalline hydrocarbon, having a bluish fluorescence, melting at 213° C. and boiling above 360° C. It was discovered in coal-tar by Dumas and Laurent in 1832, and its composition was determined by Fritzsche in 1857. It separates in the form of crystals from the anthracene oil on cooling, and is removed by filtration. The adhering oil is got rid of by submitting the crystals to great pressure in hydraulic presses. Further purification is effected by powdering the crude anthracene cake and washing with solvent naphtha, *i. e.* the mixture of the higher homologues of benzene left after the rectification of the light oil. Another coal-tar product, viz. the pyridine base referred to in the last chapter, has been recently employed for washing anthracene with great success. It is used either by itself or mixed with the solvent naphtha. The anthracene by washing with these solvents is freed from more soluble impurities, and may then contain from 30 to 80 per cent. of the pure hydrocarbon. The washing liquid, which is recovered by distillation, contains, among other impurities dissolved out of the crude anthracene, a hydrocarbon isomeric with the latter, and known as phenanthrene, for which there is at present but little use, but which may one day be turned to good account. The actual amount of anthracene contained in coal-tar corresponds to about ½lb. per ton of coal distilled, *i.e.* from ¼ to ½ per cent. by weight of the tar. Owing to the great value of alizarin and the large quantity of this colouring-matter annually consumed, anthracene is now by far the most important of the coal-tar hydrocarbons.

Alizarin, purpurin, and the other colouring-matters of madder are hydroxyl derivatives of a compound derived from anthracene by the replacement of two atoms of hydrogen by two atoms of oxygen. These oxygen derivatives of benzenoid hydrocarbons form a special group of compounds known as quinones. Thus there is quinone itself, or benzoquinone, which is benzene with two atoms of oxygen replacing two atoms of hydrogen. There are also isomeric quinones of the naphthalene series known as naphthaquinones. A dihydroxyl derivative of one of the latter is in use under the somewhat misappropriate name of "alizarin black." With this exception no other quinone derivative is used in the colour industry till we come to the hydrocarbons of the anthracene oil. Phenanthrene forms a quinone which has been utilized as a source of colouring-matters, but these are comparatively unimportant. The quinone with which we are at present concerned is anthraquinone.

The latter is prepared by oxidizing the anthracene—previously reduced by sublimation to the condition of a very finely-divided crystalline powder—with sulphuric acid and potassium dichromate. The quinone is purified, converted into a sulpho-acid, and the sodium salt of the latter on fusion with alkali gives alizarin, which is dihydroxyanthraquinone. It is of interest to note that in this case a monosulpho-acid gives a dihydroxyderivative. During the process of fusion potassium chlorate is added, by which means the yield of alizarin is considerably increased. In the original process of Graebe and Liebermann, dibromanthraquinone was fused with alkali; but this method was soon improved upon by the discovery of the sulpho-acid by Caro and Perkin in 1869, and from this period the manufacture of artificial alizarin became commercially successful.

In addition to alizarin, other anthracene derivatives are of industrial importance. The purpurin, discovered among the colouring-matters of madder in 1826, is a trihydroxy-anthraquinone; it can be prepared by the oxidation of alizarin, as shown by De Lalande in 1874. Isomeric compounds known as "flavopurpurin" and "anthrapurpurin" are also made from the disulpho-acids of anthraquinone by fusion with alkali and potassium chlorate. These two disulpho-acids are obtained simultaneously with the monosulpho-acid by the action of fuming sulphuric acid on the quinone, and are separated by the fractional crystallization of their sodium salts from the monosulpho-acid (which gives alizarin) and from each other. The purpurins give somewhat yellower shades than alizarin. Another trihydroxy-anthraquinone, although not obtained directly from anthracene, must be claimed as a tar-product. It is prepared by heating gallic acid with benzoic and sulphuric acids, or with phthalic anhydride and zinc chloride, and is a brown dye known as "anthragallol" or "anthracene-brown." The anthracene derivative is in this process built up synthetically. A sulpho-acid of alizarin has been introduced for wool dyeing under the name of alizarin carmine, and a nitro-alizarin under the name of alizarin orange. The latter on heating with glycerin and sulphuric acid is transformed into a remarkably fast colouring-matter known as alizarin blue, which is used for dyeing and printing. By heating alizarin blue with strong sulphuric acid, it is converted into alizarin green.

The great industry arising out of the laboratory work of two German chemists has influenced other branches of chemical manufacture, and has reacted upon the coal-tar colour industry itself. A new application for caustic soda and potassium chlorate necessitated an increased production of these materials. The first demand for fuming sulphuric acid on a large scale was created by the alizarin manufacture in 1873, when it was found that an acid of this strength gave better results in the preparation of sulpho-acids from anthraquinone. The introduction of this acid into commerce no doubt exerted a marked influence on the production of other valuable sulpho-acids, such as acid magenta in 1877, acid yellow in 1878, and acid naphthol yellow in 1879. The introduction of artificial alizarin has also simplified the art of colour printing on cotton fabrics to such an extent that other colouring-matters, also derived from coal-tar, are largely used in combination with the alizarin to produce parti-coloured designs. The manufacture of one coal-tar colouring-matter has thus assisted in the consumption of others.

Artificial alizarin is used in the form of a paste, which consists of the colouring-matter precipitated from its alkaline solution by acid, and mixed with water so as to form a mixture containing from 10 to 20 per cent. of alizarin. The magnitude of the industry will be gathered from the estimate that the whole quantity of anthracene annually made into alizarin corresponds to a daily production of about 65 tons of 10 per cent. paste, of which only about one-eighth is made in this country, the remainder being manufactured on the Continent. The total production of alizarin corresponds in money value to about £2,000,000 per annum. One pound of dry alizarin has the tinctorial power of 90 pounds of madder. Seeing therefore that the raw material anthracene was at one time a waste product, and that the quantity of alizarin produced in the factory corresponds to nearly five pounds of 20 per cent. paste for one pound of anthracene, it is not surprising that the artificial has been enabled to compete successfully with the natural product.

The industrial history of anthracene is thus summarized. (See opposite.)

The black, viscid residue left in the tar-still after the removal of the anthracene oil is the substance known familiarly as pitch. From the latter, after removal of all the volatile constituents, there is prepared asphalte, which is a solution of the pitchy residue in the heavy tar-oils from which all the materials used in the colour industry have been removed. Asphalte is used for varnish-making, in the construction of hard pavements, and for other purposes. A considerable quantity of pitch is used in an industry which originated in France in 1832,

AND WHAT WE GET FROM IT. 177

and which is still carried out on a large scale in that country, and to a smaller extent in this and

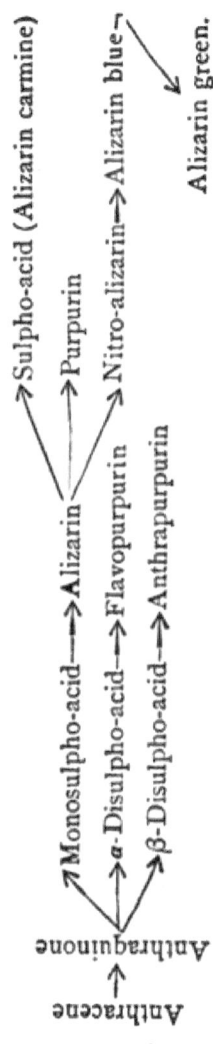

other tar-producing countries. The industry in question is the manufacture of fuel from coal-dust by moulding the latter in suitable machines with pitch so as to form the cakes known as "briquettes" or "patent fuel." By this means two waste materials are disposed of in a useful way— the pitch and the finely-divided coal, which could not conveniently be used as fuel by itself. From two to three million tons of this artificial fuel are being made annually here and on the Continent.

The various constituents of coal-tar have now been made to tell their story, so far as relates to the colouring-matters which they furnish. If the descriptive details are devoid of romance to the general reader, the results achieved in the short period of thirty-five years, dating from the discovery of mauve by Perkin, will assuredly be regarded as falling but little short of the marvellous. Although the most striking developments are naturally connected with the colouring-matters, whose history has been sketched in the foregoing pages, and whose introduction has revolutionized the whole art of dyeing, there are other directions in which the coal-tar industry has in recent times been undergoing extension. Certain tar-products are now rendering good service in pharmacy. Salicylic acid and its salts have long been used in medicine. By distilling a mixture of the dry lime salts of benzoic and acetic acids there is obtained a compound known to chemists as acetophenone, which is used for inducing sleep under the name of hypnone. The acetyl-derivative of aniline and of methylaniline are febrifuges known as "antifebrine" and "exalgine." Ethers of salicylic acid and its homologues, prepared from these acids and phenol, naphthols, &c., are in use as antiseptics under the general designation of "salols."

In 1881 there was introduced into medicine the first of a group of antipyretics derived from coal-tar bases of the pyridine series. It has already been explained that this base is removed from the light oil by washing with acid. Chemically considered, it is benzene containing one atom of nitrogen in place of a group consisting of an atom of carbon and an atom of hydrogen. The quantity of pyridine present in coal-tar is very small, and no use has as yet been found for it excepting as a solvent for washing anthracene or for rendering the alcohol used for manufacturing purposes undrinkable, as is done in this country by mixing in crude wood-spirit so as to form methylated spirit. The salts of pyridine were shown by McKendrick and Dewar to act as febrifuges in 1881, but they have not hitherto found their way into pharmacy. The chief interest of the base for us centres in the fact that it is the type of a group of bases related to each other in the same way as the coal-tar hydrocarbons. Thus in coal-tar, in addition to pyridine, there is another base known as quinoline, which is related to pyridine in the same way that naphthalene is related to benzene. Similarly there is a coal-tar base known as acridine, which is found associated with the anthracene, and which is related to quinoline in the same way that anthracene is related to naphthalene. The three hydrocarbons are comparable with the three bases, which may be regarded as derived from them in the same manner that pyridine is derived from benzene—

Benzene	Pyridine
Naphthalene	Quinoline
Anthracene	Acridine

Some of these bases and their homologues are found in the evil-smelling oil produced by the destructive distillation of bones (Dippel's oil, or bone oil), and the group is frequently spoken of as the pyridine group. The colouring-matter described as phosphine, obtained as a by-product in the manufacture of magenta (p. 94), is a derivative of acridine, and a yellow colouring-matter discovered by Rudolph in 1881, and obtained by heating the acetyl derivative of aniline with zinc chloride, is a derivative of a homologue of quinoline. This dye-stuff, known as "flavaniline," is no longer made; but it is interesting as having led to the discovery of the constitution of phosphine by O. Fischer and Körner in 1884.

The antipyretic medicines which we have first to consider are derivatives of quinoline. This base was discovered in coal-tar by Runge in 1834, and was obtained by Gerhardt in 1842 by distilling cinchonine, one of the cinchona alkaloïds, with alkali. Now it is of interest to note that the quinoline of coal-tar is of no more use to the technologist than the aniline; these bases are not contained in the tar in sufficient quantity to enable them to be separated and purified with economical advantage. If the colour industry had to depend upon this source of aniline only, its development would have been impossible. But as chemistry enabled the manufacturer to obtain aniline in quantity from benzene, so science has placed quinoline at his disposal. This important discovery was made in 1880 by the Dutch chemist Skraup, who found that by heating aniline with sulphuric acid and glycerin in the presence of nitrobenzene, quinoline is produced. The nitrobenzene acts only as an oxidizing agent; the amido-group of the aniline is converted into a group containing carbon, hydrogen, and nitrogen, *i.e.* the pyridine group. The discovery of Skraup's method formed the starting-point of a series of syntheses, which resulted in the formation of many products of technical value. In all these syntheses the fundamental change is the same, viz. the conversion of an amidic into a pyridine group. We may speak of the amido-group as being "pyridised" in such processes. Thus alizarin blue, which is formed by heating nitro-alizarin with glycerin and sulphuric acid, results from the pyridisation of the nitro-group. By an analogous method Doebner and v. Miller prepared a homologue of quinoline (quinaldine) in 1881, by the action of sulphuric acid and a certain modification of aldehyde known as paraldehyde on aniline.

Quinoline and its homologue quinaldine have been utilized as sources of colouring-matters. A green dye-stuff, known as quinoline green, was formerly made by the same method as that employed for producing the phosgene colours by Caro and Kern's process (p. 106). The phthaleïn of quinaldine was introduced by E. Jacobsen in 1882 under the name of quinoline yellow, a colouring-matter which forms a soluble sulpho-acid by the action of sulphuric acid.

To return to coal-tar pharmaceutical preparations. At the present time seven distinct derivatives of quinoline, all formed by pyridising the amido-group in aniline, amido-phenols, &c., are known in medicine under such names as kairine, kairoline, thalline, and thermifugine. The mode of preparation of these compounds cannot be entered into here. Kairine, the first of the artificial alkaloïds, is a derivative of hydroxy-quinoline, which was discovered in 1881 by Otto Fischer. All these quinoline derivatives have the property of lowering the temperature of the body in certain kinds of fevers, and may therefore be considered as the first artificial products coming into competition with the natural alkaloïd, quinine. There is reason for believing that the latter alkaloïd, the most valuable of all febrifuges, is related to the quinoline bases, so that if its synthesis is accomplished—as may certainly be anticipated—we shall have to look to coal-tar as a source of the raw materials.

Another valuable artificial alkaloïd, discovered in 1883 by Ludwig Knorr, claims aniline as a point of departure. When aniline and analogous bases are diazotised, and the diazo-salts reduced in the cold with a very gentle reducing agent, such as stannous chloride, there are formed certain basic compounds, containing one atom of nitrogen and one atom of hydrogen more than the original base. These bases were discovered in 1876 by Emil Fischer, and they are known as hydrazines, the particular compound thus obtained from aniline being phenylhydrazine. By the action of this base on a certain compound ether derived from acetic acid, which is known as aceto-acetic ether, there is formed a product termed "pyrazole," and this on methylation gives the alkaloïd in question, which is now well known in pharmacy under the name of "antipyrine."

While dealing with this first industrial application of a hydrazine, it must be mentioned that the original process by which Fischer prepared these bases was improved upon by Victor Meyer and Lecco in 1883, who discovered the use of a cold solution of stannous chloride for reducing the diazo-chloride to the hydrazine. By this method the manufacture of phenylhydrazine and other hydrazines is effected on a large scale—all kinds of amido-compounds and their sulpho-acids can be diazotised and reduced to their hydrazines. Out of this discovery has arisen the manufacture of a new class of colouring-matters related to the azo-dyes. The hydrazines combine with quinones and analogous compounds with the elimination of water, the oxygen coming from the quinone, and the hydrogen from the hydrazine. The resulting products are coloured compounds very similar in properties to the azo-dyes, and one of these was introduced in 1885 by Ziegler, under the name of "tartrazine." The latter is obtained by the action of a sulpho-acid of phenylhydrazine on dioxy-tartaric acid, and is a yellow dye, which is of special interest on account of its extraordinary fastness towards light.

Another direction in which coal-tar products

have been utilized is in the formation of certain aromatic compounds which occur in the vegetable kingdom. Thus the artificial production of bitter-almond oil from toluene has already been explained. By heating phenol with caustic alkali and chloroform, the aldehyde of salicylic acid, *i.e.* salicylic aldehyde, is formed, and this, on heating with dry sodium acetate and acetic anhydride, passes into *coumarin*, the fragrant crystalline substance which is contained in the Tonka bean and the sweet-scented woodruff. Furthermore, the familiar flavour and scent of the vanilla bean, which is due to a crystalline substance known as vanillin, can be obtained from coal-tar without the use of the plant. The researches of Tiemann and Haarman having shown that vanillin is a derivative of benzene containing the aldehyde group, one hydroxyl- and one methoxy-group, the synthesis of this compound soon followed (Ulrich, 1884). The starting-point in this synthesis is nitrobenzoic aldehyde, so that here again we begin with toluene as a raw material. A mixture of vanillin and benzoic aldehyde when attenuated to a state of extreme dilution in a spirituous solvent, gives the perfume known as " heliotrope."

Not the least romantic chapter of coal-tar chemistry is this production of fragrant perfumes from the evil-smelling tar. Be it remembered that these products—which Nature elaborates by obscure physiological processes in the living plant—are no more contained in the tar than are the hundreds of colouring-matters which have been prepared from this same source. It is by chemical skill that these compounds have been built up from their elemental groups ; and the artificial products, as in the case of indigo and alizarin, are chemically identical with those obtained from the plant. Among the late achievements in the synthesis of vegetable products from coal-tar compounds is that of juglone, a crystalline substance found in walnut-shell. It was shown by Bernthsen in 1884 that this compound was hydroxy-naphthaquinone, and in 1887 its synthesis from naphthalene was accomplished by this same chemist in conjunction with Dr. Semper.

Another recent development in the present branch of chemistry brings a coal-tar product into competition with sugar. In 1879 Dr. Fahlberg discovered a certain derivative of toluene which possessed an intensely sweet taste. By 1884 the manufacture of this product had been improved to a sufficient extent to enable it to be introduced into commerce as a flavouring material in cases where sweetness is wanted without the use of sugar, such as in the food of diabetic patients. Under the name of "saccharin," Fahlberg thus gave to commerce a substance having more than three hundred times the sweetening power of cane-sugar—a substance not only possessed of an intense taste, but not acted upon by ferments, and possessing distinctly antiseptic properties. The future of coal-tar saccharin has yet to be developed; but its advantages are so numerous that it cannot fail to become sooner or later one of the most important of coal-tar products. In cases where sweetening is required without the possibility of the subsequent formation of alcohol by fermentation, saccharin has been used with great success, especially in the manufacture of aërated waters. Its value in medicine has been recognized by its recent admission into the Pharmacopœia.

The remarkable achievements of modern chemistry in connection with coal-tar products do not end with the formation of colouring-matters, medicines, and perfumes. The introduction of the beautiful dyes has had an influence in other directions, and has led to results quite unsuspected until the restless spirit of investigation opened out new fields for their application. A few of these secondary uses are sufficiently important to be chronicled here. In sanitary engineering, for example, the intense colouring power of fluorescein is frequently made use of to test the soundness of drains, or to find out whether a well receives drainage from insanitary sources. In photography also coal-tar colouring-matters are playing an important part by virtue of a certain property which some of these compounds possess.

The ordinary photographic plate is, as is well known, much more sensitive to blue and violet than to yellow or red, so that in photographing coloured objects the picture gives a false impression of colour intensity, the violets and blues impressing themselves too strongly, and the yellows and reds too feebly. It was discovered by Dr. H. W. Vogel in 1873 that if the sensitive film is slightly tinted with certain colouring-matters, the sensitiveness for yellow and red can be much increased, so that the picture is a more natural representation of the object. Plates thus dyed are said to be "isochromatic" or "orthochromatic," and by their use paintings or other coloured objects can be photographed with much better results than by the use of ordinary plates. The boon thus conferred upon photographic art is therefore to be attributed to coal-tar chemistry. Among the numerous colouring-matters which have been experimented with, the most effective special sensitizers are erythrosin, one of the phthaleïns, quinoline red, a compound related to the same group, and cyanin, a fugitive blue colouring-matter obtained from quinoline in 1860 by Greville Williams.

In yet another way has photography become indebted to the tar chemist. Two important developers now in common use are coal-tar products, viz. hydroquinone and eikonogen. The history of these compounds is worthy of narration as showing how a product when once given by chemistry to the world may become applicable in quite unexpected directions. Chloroform is a case in point. This compound was discovered by Liebig in 1831, but its use as an anæsthetic did not come about till seventeen years after its discovery. It was Sir James Simpson who in 1848 first showed the value of chloroform in surgical operations. A similar story can be told with respect to these photographic developers.

Towards the middle of the last century a French chemist, the Count de la Garaye, noticed a crystalline substance deposited from the extract of Peruvian bark, then, as now, used in medicine. This substance was the lime salt of an acid to which Vauquelin in 1806 gave the name of quinic acid (*acide quinique*). In 1838 Woskresensky, by oxidizing quinic acid with sulphuric acid and oxide of manganese, obtained a crystalline substance which he called quinoyl. The name was changed to quinone by Wöhler, and, as we have already seen (p. 172), the term has now become generic, indicating a group of similarly constituted oxygen derivatives of hydrocarbons. Hydroquinone was obtained by Caventou and Pelletier by heating quinic acid, but these chemists did not recognize its true nature. It was the illustrious Wöhler who in 1844 first prepared the compound in a state of purity, and established its relationship to quinone. This relationship, as the name given by Wöhler indicates, is that of the nature of a hydrogenised quinone. The compound is readily prepared by the action of sulphurous acid or any other reducing agent on the quinone.

It has long been known in photography, that a developer must be of the nature of a reducing agent, either inorganic or organic, and many hydroxylic and amidic derivatives of hydrocarbons come under this category. Thus, pyrogallol, which has already been referred to as a trihydroxybenzene (p. 146), when dissolved in alkali rapidly absorbs oxygen—it is a strong reducing agent, and is thus of value as a developer. But although pyrogallol is a benzene derivative, and could if necessary be prepared synthetically, it can hardly be claimed as a tar product, as it is generally made from gallic acid. Now hydroquinone when dissolved in alkali also acts as a reducing agent, and in this we have the first application of a true coal-tar product as a photographic developer. Its use for this purpose was suggested by Captain Abney in 1880, and it was found to possess certain advantages which caused it to become generally adopted.

As soon as a practical use is found for a chemical product its manufacture follows as a matter of course. In the case of hydroquinone, the original source, quinic acid, was obviously out of question, for economical reasons. In 1877, however, Nietzki worked out a very good process for the preparation of quinone from aniline by oxidation with sulphuric acid and bichromate of soda in the cold. This placed the production of quinone on a manufacturing basis, so that when a demand for hydroquinone sprung up, the wants of the photographer were met by the technologist. Eikonogen is another organic reducing agent, discovered by the writer in 1880, and introduced as a developer by Dr. Andresen in 1889. It is an amido-derivative of a sulpho-acid of beta-naphthol, so that naphthalene is the generating hydrocarbon of this substance.

The thio-derivative of toluidine described as " primuline " (p. 160), has recently been found by its discoverer to possess a most remarkable property which enables this compound to be used for the photographic reproduction of designs in azo-colours. Diazotised primuline, as already explained, combines in the usual way with amines and phenols to form azo-dyes. Under the influence of light, however, the diazotised primuline is decomposed with the loss of nitrogen, and the formation of a product which does not possess the properties of a diazo-compound. The product of photochemical decomposition no longer forms azo-colours with amines or phenols. If, therefore, a fabric is dyed with primuline, then diazotised by immersion in a nitrite bath, and exposed under a photographic negative, those portions of the surface to which the light penetrates lose the power of giving a colour with amines or phenols. The design can thus be developed by dipping the fabric into a solution of naphthol, naphthylamine, &c. By this discovery another point of contact has been established between photography and coal-tar products. Nor is this the only instance of its kind, for it has also been observed that a diazo-sulpho-acid of one of the xylenes does not combine with phenols to form azo-dyes excepting under the influence of light. A fabric can therefore be impregnated with the mixture of diazo-sulpho-acid and naphthol, and exposed under a design, when the azo-colour is developed only on those portions of the surface which are acted upon by light.

The last indirect application of coal-tar colouring-matters to which attention must be called is one of great importance in biology. The use of these dyes as stains for sections of animal and vegetable tissue has long been familiar to microscopists.

Owing to the different affinities of the various components of the tissue for the different colouring-matters, these components are capable of being differentiated and distinguished by microscopical analysis. Furthermore, the almost invisible organisms which in recent times have been shown to play such an important part in diseases, have in many cases a special affinity for particular colouring-matters, and their presence has been revealed by this means. The micro-organism of tubercle, for example, was in this way found by Koch to be readily stained by methylene blue, and its detection was thus rendered possible with certainty. Many of the dyes referred to in the previous pages have rendered service in a similar way. To the pure utilitarian such an application of coal-tar products will no doubt compensate for any defects which they may be supposed to possess from the æsthetic point of view.[1]

From a small beginning there has thus developed in a period of five-and-thirty years an enormous industry, the actual value of which at the present time it is very difficult to estimate. We shall not be far out if we put down the value of the coal-tar

---

[1] Since the above was written the continuation of Koch's researches upon the tubercle bacillus has culminated in the discovery of his now world-renowned lymph for the inoculation of patients suffering from tubercular disease.

colouring-matters produced annually in this country and on the Continent at £5,000,000 sterling. The products which half a century or so ago were made in the laboratory with great difficulty, and only in very small quantities, are now turned out by the hundredweight and the ton.[1] To achieve these results the most profound chemical knowledge has been combined with the highest technological skill. The outcome has been to place at the service of man, from the waste products of the gas-manufacturer, a series of colouring-matters which can compete with the natural dyes, and which in many cases have displaced the latter. From this source we have also been provided with explosives such as picric acid; with perfumes and flavouring materials like bitter-almond oil and vanillin; with a sweetening principle like saccharin—compared with which the product of the sugar-cane is but feeble; with dyes which tint the photographic film, and enable the most delicate gradations of shade to be reproduced; with developers such as hydroquinone and eikonogen; with disinfectants which contribute to the healthiness of our towns; with potent medicines which rival the natural alkaloïds;

---

[1] In one large factory in Yorkshire there is a set of stills kept constantly at work making pure aniline at the rate of two hundred tons per month. The monthly consumption of coal in this factory is two thousand tons, equal to twenty-four thousand tons per annum.

and with stains which reveal the innermost structure of the tissues of living things, or which bring to light the hidden source of disease. Surely if ever a romance was woven out of prosaic material it has been this industrial development of modern chemistry.

But although the results are striking enough when thus summed up, and although the industrial importance of all this work will be conceded by those who have the welfare of the country in mind, the paths which the pioneers have had to beat out can unfortunately be followed but by the few. It is not given to our science to strike the public mind at once with the magnitude of its achievements, as is the case with the great works of the engineer. Nevertheless the scientific skill which enables a Forth Bridge to be constructed for the use of the travelling public of this age—marvellous as it may appear to the uninstructed—is equalled, if not surpassed, by the mastery of the intricate atomic groupings which has enabled the chemist to build up the colouring-matters of the madder and indigo plants.

A great industry needs no excuse for its existence provided that it supplies something of use to man, and finds employment for many hands. The coal-tar industry fulfils these conditions, as will be gathered from the foregoing pages. If any further justification is required from a more exalted standpoint than that of pure utilitarianism it can be supplied. It is well known to all who have traced the results of applying any scientific discovery to industrial purposes, that the practical application invariably reacts upon the pure science to the lasting benefit of both. In no department of applied science is this truth more forcibly illustrated than in the branch of technology of which I have here attempted to give a popular account. The pure theory of chemical structure—the guiding spirit of the modern science—has been advanced enormously by means of the materials supplied by and resulting from the coal-tar industry. The fundamental notion of the structure of the benzene molecule marks an epoch in the history of chemical theory of which the importance cannot be too highly estimated. This idea occurred, as by inspiration, to August Kekulé of Bonn in the year 1865, and its introduction has been marked by a quarter century of activity in research such as the science of chemistry has never experienced at any previous period of its history. The theory of the atomic structure of the benzene molecule has been extended and applied to all analogous compounds, and it is in coal-tar that we have the most prolific source of the compounds of this class.

It was scarcely to be wondered at that an idea

which has been so prolific as a stimulator of original investigation should have exerted a marked influence on the manufacture of tar-products. All the brilliant syntheses of colouring-matters effected of late years are living witnesses of the fertility of Kekulé's conception. In the spring of 1890 there was held in Berlin a jubilee meeting commemorating the twenty-fifth anniversary of the benzene theory. At that meeting the representative of the German coal-tar colour industry publicly declared that the prosperity of Germany in this branch of manufacture was primarily due to this theoretical notion. But if the development of the industry has been thus advanced by the theory, it is no less true that the latter has been helped forward by the industry.

The verification of a chemical theory necessitates investigations for which supplies of the requisite materials must be forthcoming. Inasmuch as the very materials wanted were separated from coal-tar and purified on a large scale for manufacturing purposes, the science was not long kept waiting. The laborious series of operations which the chemist working on a laboratory scale had to go through in order to obtain raw materials, could be dispensed with when products which were at one time regarded as rare curiosities became available by the hundredweight. It is perhaps not too much to say that the advancement of chemical theory in the direction started by Kekulé has been accelerated by a century owing to the circumstance that coal-tar products have become the property of the technologist. In other words, we might have had to wait till 1965 to reach our present state of knowledge concerning the theory of benzenoid compounds if the coal-tar industry had not been in existence. And this is not the only way in which the industry has helped the science, for in the course of manufacture many new compounds and many new chemical transformations have been incidentally discovered, which have thrown great light on chemical theory. From the higher standpoint of pure science, the industry has therefore deservedly won a most exalted position.

With respect to the value of the coal-tar dyes as tinctorial agents, there is a certain amount of misconception which it is desirable to remove. There is a widely-spread idea that these colours are fugitive—that they rub off, that they fade on exposure to light, that they wash out, and, in short, that they are in every way inferior to the old wood or vegetable dyes. These charges are unfounded. One of the best refutations is, that two of the oldest and fastest of natural colouring-matters, viz. alizarin and indigo, are coal-tar products. There are some coal-tar dyes which are not fast to light, and there are many vegetable dyes which are equally fugitive. If there are natural colouring-matters which are fast and which are æsthetically orthodox, these are rivalled by tar-products which fulfil the same conditions. Such dyes as aniline black, alizarin blue, anthracene brown, tartrazine, some of the azo-reds and naphthol green resist the influence of light as well as, if not better than, any natural colouring-matter. The artificial yellow dyes are as a whole faster than the natural yellows. There are at the present time some three hundred coal-tar colouring-matters made, and about one-tenth of that number of natural dyes are in use. Of the latter only ten—let us say 33 per cent.—are really fast. Of the artificial dyes, thirty are extremely fast, and thirty fast enough for all practical requirements, so that the fast natural colours have been largely outnumbered by the artificial ones. If Nature has been beaten, however, this has been rendered possible only by taking advantage of Nature's own resources—by studying the chemical properties of atoms, and giving scope to the play of the internal forces which they inherently possess—

> " Yet Nature is made better by no mean,
> But Nature makes that mean : so, o'er that art,
> Which, you say, adds to Nature, is an art
> That Nature makes."

COAL;

The story told in this chapter is chronologically summarized below—

1820. Naphthalene discovered in coal-tar by Garden.
1832. Anthracene discovered in coal-tar by Dumas and Laurent.
1834. Phenol discovered in coal-tar by Runge.
1842. Picric acid prepared from phenol by Laurent; manufactured in Manchester in 1862.
1845. Benzidine discovered by Zinin.
1859. Corallin and aurin discovered by Kolbe and Schmitt and by Persoz; leading to manufacture from oxalic acid and phenol.
1860. Synthesis of salicylic acid by Kolbe.
1864. Manchester yellow discovered by Martius, leading to manufacture of alpha-naphthylamine and then to alpha-naphthol.
1867. Magdala red discovered by Schiendl.
1868. Synthesis of alizarin by Graebe and Liebermann, leading to the utilization of anthracene, caustic soda, potassium chlorate and bichromate, and calling into existence the manufacture of fuming sulphuric acid.
1870. Galleïn, the first of the phthaleïns, discovered by A. v. Baeyer, followed in 1871 by cœruleïn, and in 1874 by the eosin dyes (Caro). These discoveries necessitated the manufacture of phthalic acid and resorcinol.
1873. Orthochromatic photography discovered by Vogel.
1876. Azo-dyes from the naphthols introduced by Roussin and Poirrier and Witt, leading to the manufacture of the naphthols, sulphanilic acid, &c.
1877. Preparation of quinone from aniline by Nietzki,

utilized in photography in 1880 for manufacture of hydroquinone.

1878. Disulpho-acids of beta-naphthol introduced by Meister, Lucius, and Brüning, leading to azo-dyes from aniline, toluidine, xylidine, and cumidine.

1879. Acid naphthol yellow introduced by Caro.

,, Biebrich scarlet, the first secondary azo-colour, introduced by Nietzki.

,, Nitroso-sulpho acid of beta-naphthol discovered by the writer; followed in 1883 by naphthol green (O. Hoffmann), and in 1889 by eikonogen (Andresen).

,, Beta-naphthol violet, the first of the oxazines, discovered by the writer; followed in 1881 by gallocyanin.

,, Coal-tar saccharin discovered by Fahlberg; manufacture made practicable in 1884.

1880. Synthesis of indigo by A. v. Baeyer.

,, Quinoline synthesised by Skraup's process.

1881. Kairine introduced by O. Fischer, the first artificial febrifuge.

,, Indophenol discovered by Köchlin and Witt.

,, Azo-dyes from new sulpho-acid of beta-naphthol introduced by Bayer & Co.

1883. Antipyrine introduced by L. Knorr, leading to manufacture of phenylhydrazine.

1884. Congo red, the first secondary azo-colour from benzidine, introduced by Böttiger. Beginning of manufacture of cotton azo-dyes, and leading to the production of benzidine and tolidine on a large scale.

1885. Secondary azo-dyes from benzidine and tolidine containing two dissimilar amines, phenols, &c., introduced by Pfaff.

,, Tartrazine discovered by Ziegler; manufacture of

sulpho-acid of phenylhydrazine and of dioxytartaric acid.

1885. Thiorubin introduced by Dahl & Co., leading to manufacture of thiotoluidine; followed by primuline, discovered by A. G. Green in 1887.

1886. Secondary azo-dyes of stilbene series introduced by Leonhardt & Co.

ADDENDUM.

By passing steam over red-hot carbon, a mixture of carbon monoxide and hydrogen is formed. This mixture of inflammable gases is known as "water-gas," and in the preparation of the gas on a large scale, coke is used as a source of carbon. If, therefore, water-gas became generally used, another use for coke would be added to those already referred to (p. 47).

With reference to the consumption of coal in London (p. 46), it appears from the Report of a Committee of the Corporation of London, issued at the end of 1890, that the present rate of consumption in the Metropolis is 9,709,000 tons per annum. This corresponds to 26,600 tons per diem. It has been proved by experiment, that when coal is burnt in an open grate, from one to three per cent. of the coal escapes in the form of unburnt solid particles, or "soot," and about 10 per cent. is lost in the form of volatile compounds of carbon. It has been estimated that the total amount of coal annually wasted by imperfect combustion in this country is 45,000,000 tons, corresponding to about £12,000,000, taking the value of coal at the pit's mouth. Taking the unconsumed solid particles at the very lowest estimate of 1 per cent., it will be seen that, in London alone, we are sending forth carbonaceous and tarry matter into the atmosphere at the rate of about 266 tons daily; and volatile carbon compounds at the daily rate of 2660 tons (see p. 32). At the price of coal in London this means that, in solid combustibles alone, we are absolutely squandering about £10,000 annually, to say nothing of the damage caused by the presence of this sooty pall. Such facts as these require no comment; they speak for themselves in sombre gloom, and in the sickliness of our town vegetation—they give a new meaning to the term "in darkest London," and they plead eloquently for science and legislation to show us "the way out."

INDEX.

Accum, condensation of tar-oils, 69
Acetic acid, 64
Acetophenone, 178
Acid brown, 151
Acid greens, 106
Acid magenta, 92
Acid naphthol yellow, 143
Acid yellow, 120
Acridine, 180
Air, composition of, 24
Albo-carbon light, 140
Alizarin black, 172
Alizarin blue, 174
Alizarin carmine, 174
Alizarin green, 174
Alizarin orange, 174
Alkali blue, 93
Amidodimethylaniline, 112
Ammonia in gas-liquor, 64
Ammonia, origin in gas-liquor, 65
Analyses of coal, 23
Aniline black, 114
Aniline, history, 75
Aniline, manufacture, 87
Aniline yellow, 116
Annual production of ammonia, 68
Anthracene, 171
Anthracene brown, 174
Anthracene oil, 81, 167
Anthragallol, 174

Anthrapurpurin, 174
Anthraquinone, 173
Antifebrine, 179
Antipyrine, 183
Archil substitute, 151
Arctic coal, 12
Arsenic acid process, 90
Arsenic acid, recovery, 94
Artificial alizarin, 173
Artificial purpurin, 173
Asphalte, 176
Auramine, 106
Aurin, 132
Azines, 109
Azo-blacks, 159
Azo-colours, 118
Azo-dyes for cotton, 158
Azobenzene, 119
Azo-dyes from salicylic acid, 135
Azotoluene, 119

Baeyer, A.v., indigo, 127
Baeyer, A.v., phthaleïns, 146
Basle blue, 111
Becher, early experiments, 34
Beet-sugar cultivation, 67
Benzal chloride, 102
Benzaldehyde, 103
Benzene, discovery, 73
Benzene, final purification, 86
Benzene in tar-oil, 73
Benzene theory, 196
Benzidam, 75

INDEX.

Benzidine, 136
Benzoic acid, manufacture, 104
Benzotrichloride, 102
Benzyl chloride, 102
Bernthsen, methylene blue, 113
Bethell, timber preserving, 70
Biebrich scarlet, 156
Biology, dyes used in, 192
Bismarck brown, 116
Bitter-almond oil, 103
Bone oil, 180
Böttiger, Congo-red, 157
Bréant, timber pickling, 70
Briquettes, 178
Burning naphtha, 86

CALORIFIC value of carbon, 25
Calorific value of coal, 20
Carbolic acid, 129
Carbolic oil, 23, 80, 129
Carbon dioxide, 24
Carboniferous period, 11
Caro and Kern, phosgene dyes, 106
Caro and Wanklyn, rosolic acid, 133
Caro, fluorescein and eosin, 147
Caro, methylene blue, 112
Chemical washing, 83
Chlorophyll, 29
Chrysamines, 136
Chrysoïdine, 116
Cinnamic acid, 128
Clayton, Dean, distils coal, 35
Clegg, Samuel, gas-engineer, 40
Coal, amount raised, 59
Coal-fields of United Kingdom, 60
Coal-gas, composition, 56
Coal-gas, manufacture, 42
Coal-mining, history, 57
Coal, origin, 9
Coal, supply, 59
Cœruleïn, 147

Coke, composition of, 48
Coke from gas-retorts, 45
Coke-oven tar, 49
Coke, uses of, 47
Combustion, 23
Composition of coal, 23
Congo red, 157
Conservation of energy, 18
Constitution of molecules, 95
Corallin, 132
Cotton dyes, 137
Coumarin, 185
Coupier's process, 91
Creosote oil, 163
Creosoting of timber, 70
Cresols, 131
Cresylic acid, 131
Cretaceous coal, 12
Crocein scarlets, 156
Crystal violet, 106
Cumidine, 155
Cyanin, 188

DALE and Caro, induline, 120
Destructive distillation, 33
Diazo-compounds, 116
Diazotype, 191
Dimethylaniline, 100
Dinitrobenzene, 120
Diphenylamine, 101
Diphenylamine blue, 101
Distillation, fractional, 78
Doebner, malachite green, 102
Dover, coal under, 60
Dumas and Laurent, anthracene, 171
Dundonald, Earl, early experiments, 39

EIKONOGEN, 189
Electricity as an illuminating agent, 62
Eocene coal, 12
Eosin, 147

INDEX. 207

Erythrosin, 188
Essence of mirbane, 74
Exalgine, 179

FAHLBERG, saccharin, 186
Faraday, discovers benzene, 73
Fast red, 151
Fertilization by ammonia, 66
Fire-damp, 32
First runnings, 80
Fischer, E. and O., rosaniline, 97
Fischer, hydrazines, 183
Fischer, malachite green, 103
Flavaniline, 180
Flavopurpurin, 174
Fluorescein, 147
Foot-pound, 19
Fractional distillation, 78
Fritzsche, aniline, 75

GALLEIN, 146
Gallic acid, 146
Gallocyanin, 162
Gambines, 161
Garancin, 168
Garden, naphthalene, 139
Gas producers, 57
Gas, quantity obtained from coal, 45
Girard and De Laire, rosaniline blues, 92
Glucosides, 169
Goethe, visit to coke-burner, 48
Græbe and Liebermann, alizarin, 170
Graphite, 12
Grässler, acid yellow, 120
Green, A. G., primuline, 160
Griess, diazo-compounds, 116

HALES, Rev. Stephen, distils coal, 37
Hofmann, benzene in tar-oil, 73

Hofmann, red from aniline, 89
Hofmann's violets, 93
Homologous series, 152
Horse-power, 22
Hull, Prof., coal supply, 59
Hydrazines, 183
Hydrocarbons of benzene series, 82
Hydrogen, calorific value, 25
Hydroquinone, 189
Hypnone, 179

INDIGO plants, 124
Indigo, syntheses, 126
Indophenol, 162
Indulines, 121
Ingrain colours, 160
Iodine green, 94
Iron smelting, 16
Iron swarf, 87
Isochromatic plates, 188
Isomerism, 88

JOULE, mechanical equivalent of heat, 19
Juglone, 186
Jurassic coal, 12

KAIRINE, 182
Kekulé, benzene theory, 196
Kekulé and Hidegh, azo-dyes, 135
Koch, tubercle, 193
Köchlin, gallocyanin, 162
Köchlin and Witt, indophenol, 162
Kolbe, salicylic acid, 134
Kolbe and Schmitt, phenol dye, 132
Kyanol, 75

LAMPBLACK, 139, 166
Laurent, phenol, 131
Laurent, phthalic acid, 143

Laurent, picric acid, 136
Lauth, methyl violet, 98
Lauth's violet, 111
Leonhardt & Co., stilbene dyes, 137
Lightfoot, aniline black, 114
Light oil, 80
Light oils, early uses, 71
Lister, antiseptic surgery, 131
London, coal introduced, 58
London, illuminated by gas, 40
Lucigen burner, 163

MADDER, 168
Magdala red, 149
Magenta, history, 89
Malachite green, 102
Manchester brown, 116
Manchester yellow, 142
Mansfield, isolation of benzene, 73
Mansfield's still, 77
Manures, 66
Marsh gas, 32
Mauve, discovery, 74
Mechanical value of coal, 22
Medlock, magenta process, 90
Methyl chloride, 99
Methylene green, 113
Methyl green, 101
Methyl violet, 98
Mirbane, essence, 74
Murdoch, introduces coal-gas, 40

NAPHTHALENE, annual production, 141
Naphthalene in carbolic oil, 130
Naphthionic acid, 151
Naphthol green, 161
Naphthol orange, 151
Naphthols, 141
Naphthylamines, 142
Natanson, aniline red, 89
Neutral red, 111

Neutral violet, 111
New blue, 161
Nicholson blue, 93
Nicholson, magenta process, 90
Nietzki, azines, 109
Nietzki, Biebrich scarlet, 156
Nietzki, quinone, 191
Night blue, 106
Nigrosine, 121
Nitrification, 66
Nitrobenzene process, 91
Nitrosodimethylaniline, 111
Non-Carboniferous coal, 11
Number of compounds in tar, 81

OLD Red Sandstone, coal in, 12
Oligocene brown coal, 12
Orthochromatic plates, 188
Oxazines, 162
Oxide of iron for gas purifying, 44

PARAFFIN oil and wax, 50
Patent fuel, 178
Perfumes, 185
Perkin, alizarin, 170
Perkin, discovers mauve, 74
Permanence of dyes, 198
Permian coal, 12
Persoz, phenol dye, 132
Pharmaceutical preparations, 178
Phenanthrene, 172
Phenols, 129
Phosgene dyes, 106
Phosphine, 94, 180
Photographic developers, 189
Phthaleïns, 146
Phthalic acid, 143
Picric acid, 136
Pitch, 81, 176
Plants, growth of, 26
Ponceaux, 151
Primary azo-dyes, 156
Primuline, 160

Propiolic acid, 128
Purpurin, 169
Pyrazole, 183
Pyridine, 87
Pyridine bases, 179
Pyrogallol, 146

QUINALDINE, 182
Quinic acid, 189
Quinoline, 180
Quinoline green, 182
Quinoline red, 188
Quinoline yellow, 182
Quinones, 172

READ HOLLIDAY'S lamp, 72
Rectification of hydrocarbons, 84
Resorcinol, 145
Rhodamines, 148
Roccellin, 151
Roman coal-mining, 57
Rosaniline blues, 92
Rosaniline from rosolic acid, 133
Rosolic acid, 132
Runge, kyanol, 75
Runge, phenol in tar, 131

SACCHARIN, 186
Saffranine, 108
Salicylic acid, 134
Salicylic aldehyde, 185
Salols, 179
Schiendl, Magdala red, 149
Scotland, shale-oil industry, 53
Sea coal, 58
Secondary azo-dyes, 156
Shale, nature of, 51
Shale-oil industry, 50
Skraup, quinoline, 181
Sodium nitrite, 119
Solar energy, 28
Soluble blue, 93
Solvent naphtha, 86
Steam-engines, 20

Stilbene azo-dyes, 137
Structure of coal, 15
Sulphanilic acid, 150
Sulphuric acid, 45
Sunlight, source of energy in coal, 30
Surgery, antiseptic, 131

TAR distilling, 77
Tar, first utilization, 69
Tar from coke-ovens, 49
Tar, quantity obtained from coal, 45
Tartrazine, 184
Tertiary coal, 12
Thalline, 182
Thermifugine, 182
Thiazines, 113
Thiodiphenylamine, 113
Thiorubin, 160
Timber pickling, 70
Tolidine, 136
Toluene from tar, 81
Torbane Hill coal, 52
Transformation of wood into coal, 31
Triassic coal, 12
Trimethylamine, 99
Trinitrophenol, 136
Triphenylmethane, 97
Tubercle bacillus, 193
Turkey red, 170

UNDERCLAY, 13
Unverdorben, crystallin, 75
Uses of coal, 16

VANILLIN, 185
Vegetable deposits, recent, 13
Verguin, red from aniline, 90
Victoria blue, 106
Victoria yellow, 137
Vinasse, 99
Vincent, methyl chloride, 100

O

INDEX.

Violaniline, 121

WASTEFUL use of coal, 32, 203
Water, composition of, 25
Water gas, 203
Watson, Bishop, coke-oven tar, 50
Watson, Bishop, distils coal, 37
Wealden iron industry, 17
Whitaker, W., coal in S.-E. England, 60
Winsor, promotes introduction of gas, 41.

Witt, azines, 109
Witt, chrysoïdine, 116
Witt, naphthol orange, 150
Wood vinegar, 64
Woody fibre, 26
Woulfe, picric acid, 136

XYLENES from tar, 81
Xylidine scarlet, 153

YOUNG, James, burning-oil, 51

ZIEGLER, tartrazine, 184
Zinin, benzidam, 75

THE END.

Richard Clay & Sons, Limited, London & Bungay.

www.ingramcontent.com/pod-product-compliance
Lightning Source LLC
Chambersburg PA
CBHW020832230426
43666CB00007B/1190